Published by Cool Springs Press
P.O. Box 2828
Brentwood, Tennessee 37024

EAN: 978-1-59186-451-6

First Printing 2009
Printed in the United States of America
10 9 8 7 6 5 4 3 2 1

Art Director: Marc Pewitt
Horticulture Editor: Andrew Bunting
Copyeditor: Billie Brownell

Visit the Cool Springs Press Web site at www.coolspringspress.com.

# PROVEN PLANTS
## SOUTHERN GARDENS

# ERICA GLASENER

COOL
SPRINGS
PRESS

*Growing Successful Gardeners*™

www.coolspringspress.com
BRENTWOOD, TENNESSEE

## DEDICATION

To my daughter, Georgia, who helps me appreciate the beauty and wonder of the natural world every day.

## ACKNOWLEDGEMENTS

I want to thank my family and friends for their continued support in all my ventures both professional and personal. Thanks to Walter Reeves, my co-author for two books about gardening in Georgia, for his assistance. I especially appreciate Andrew Bunting for his horticultural acumen and checking my nomenclature; my friends in the nursery business for sharing their knowledge, time, and plants with me; and the many gardeners I have met over the years who inspire me to garden.

# CONTENTS

## APPENDIX

It is my hope that this book will prove useful for Southern gardeners, of all types, especially those new to gardening in the South and those who may be overwhelmed by the many choices of plants offered in nurseries. For each category of plants that I have included, of which there are twenty, I have recommended ten plants that perform well in our Southern climate. In addition to the name of the plant, I provide some information about it, as well as suggestions on how to use it in the landscape. In some cases, I have intentionally *not* recommended plants, such as English Ivy for a groundcover. In this case, I believe English Ivy is aggressive and invasive and that there are much better alternatives for our ornamental gardens. Some of the plants I have recommended are tried-and-true, while others may not be as familiar, but are truly garden worthy. I chose plants for the twenty categories based on their best attributes. In some cases, I recommend a tree for its outstanding ornamental bark, but it may also be a choice shade tree, like the Lacebark Elm, *Ulmus parvifolia*.

If this book whets your appetite and motivates you to try plants that you haven't considered yet or didn't know existed—great! I realize that not everyone may agree with all my choices, but they are based on my years of experience in my own garden.

PATIO GARDEN

## PLANT NAMES

Throughout this book I have included both the scientific and common names for plants. While scientific names can change, this is the exception rather than the rule. Scientific names are shown written in italic, and any cultivar names are surrounded by single quotes. For example, *Acer rubrum* 'October Glory' is a selection of Red Maple called 'October Glory'. Armed with both the scientific and common names, it should be easier for you to track down a specific wonderful plant for your garden.

## PLAN BEFORE YOU PLANT

Hiring a professional garden designer or landscape architect to come up with a plan for your garden will be money well spent, especially if you are new to gardening. A landscape plan gives you something to work with, and as you become more familiar with plants and your landscape, you can tailor the plan to suit your personal taste.  A landscape plan can actually save money by avoiding those plants that the landscape designer knows will not work for our Southern climate or in your particular space.

## SOIL, SUN, AND COMPOST

A gardening phrase that is popular and useful when starting a new

garden or refurbishing an existing one is "the right plant for the right place."

I cannot stress the importance of this enough. With this approach of selecting the right plant for its place, you will be well on your way to creating a successful garden. Happy plants are more resistant to pest and disease problems and, once established in the garden, are better equipped to tolerate periods of stress, including drought. Just because a plant is a native doesn't mean that it will thrive in any garden setting. Before you plant, whether you use plants that are native to your area or ornamentals from another part of the world, take time to learn about your environment as well as the requirements of individual plants.

SHADE GARDEN

What is your exposure? The amount of sunlight that your garden receives is critical. Do you get morning sun, afternoon sun, or filtered shade all day?

**Full sun** is considered to be at least 6 to 8 hours of unfiltered sun every day.

**Part shade** can be a location in full sun all day that is filtered through a canopy of mature trees like pines, or hardwoods such as maples or oaks.

**Shade** is filtered sun all day through a dense canopy of hardwoods or direct sun that hits plants for less than 2 hours per day.

**Dense shade** means that no direct sun hits the site all day, such as a location under a Southern Magnolia.

Conducting a soil test is also a good idea and is not as complicated as it may sound. Contact your local Extension Service for more information or purchase a soil-testing kit at your local home improvement center. Amending your soil—before you plant—with compost, soil amendments, and/or organic matter helps roots absorb water and nutrients more efficiently. (Trees seem to do better if they are planted in existing soils, except for those in waterlogged soils.)

How much compost is the right amount? For flowerbeds, spread a layer of compost 2 inches thick and mix it in with the soil underneath to a depth of 6 to 8 inches. For perennials and shrubs it is a good idea to topdress (apply organic matter, compost, or composted manure around the base of shrubs, perennials, or trees) at least once per season, in very early spring or fall. Amounts vary depending upon the individual plant's requirements, but one inch is a good amount for compost. First rake away any old mulch, spread the compost, and then add fresh mulch. The compost will break down over time and continue to feed the soil.

If you use commercial fertilizers, be sure to follow the directions on the label carefully. More is not better.

## HOW TO START YOUR OWN COMPOST PILE

Bagged compost is readily available but you may want to start your own compost pile. Compost is full of tiny fungi, bacteria, and other creatures that help break down grass clippings, leaves, vegetables, and other organic materials in your compost pile. This material is then ultimately available to help make the soil soft and loose (a term known as "friable"). To start a compost pile, the basic rule is "pile it up and let it rot." For more specific information on how to start a compost pile visit www.compostguide.com.

## PLANTING

Just a brief word about planting: container-grown plants are easy to transplant to the garden. Once you prepare the soil, dig a hole that is at least as deep and twice as wide as the container the plant is

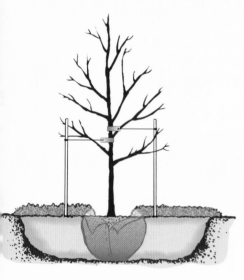

growing in. Remove the plant from its pot and trim away any obviously dead roots before planting. You can add water to the hole, let the soil settle, and then add more soil. Do not pile too much soil on top of the existing roots—plant roots need air circulation, which is why compost is so valuable. Lightly tamp the soil down around the roots until it is level with the top of the hole. Just lightly cover the top layer; don't mound up the soil. After planting, apply a 1-inch layer of mulch, keeping any mulch away from the center stems or trunks.

## WATER AND DROUGHT

Of the plants I have recommended, many are "drought friendly." What this means is that once they become established (one or two growing seasons), they can tolerate periods without any rainfall. However, it is important to remember that new plantings need water

ROSE ARBOR

on a regular basis. Many areas of the South have experienced

extended periods of drought and water restrictions over the past

five years or even longer. While this may not always be the case and

some areas may experience too much rain, it is worth thinking about

when you garden. If you don't already have them, rain barrels or

cisterns (these can be aboveground or underground) are a great way

to collect and store rainwater. You'd be amazed at how quickly they

will fill up. Of course, they are quick to empty, too. There are many

resources for both of these; check the Internet.

Mulching is also important for protecting plants, roots and retaining moisture. There are many choices for mulch; check with your local retail home and garden shop.

## SOME GUIDELINES FOR WATERING

Water container plantings as needed, when the soil is dry to the touch one inch down. Water until the water runs out of the bottom of the container. Repeat and then let the soil dry out before watering again.

### NEW PLANTINGS

Use a hose, hand-held or soaker depending on the restrictions in your area, to thoroughly soak the root ball once a week. As plants grow, remember that the root ball grows. Water accordingly. With smaller plants, annuals and perennials, you may need to water twice a week while they become established. Sandy soils may require more frequent watering while those with high clay content may require less. A water wand, which attaches to the end of a hose, is an invaluable tool.

### MATURE PLANTINGS

During an extended drought, watering your mature trees will be a wise use of your time and resources. Remember that a tree's roots extend as far as the tree's canopy. Long and slow is the rule to follow. A guideline is 600 gallons per 1000 square feet, once a week. For this type of job, an inexpensive water timer is a good investment.

### NUTRIENTS AND FERTILIZER

The most important nutrients for healthy plant growth are nitrogen, phosphorous, and potassium. When you buy a bag of fertilizer you will see three numbers representing the N, P, K on the label. These numbers represent the percentage of nitrogen, phosphorus, and potassium in the mixture. For example, a common ratio is 10-10-10, meaning 10 percent is nitrogen, 10 percent is phosphorus, and 10 percent is potassium. The remaining 70 percent is clay. Fertilizers also include micronutrients.

**The letters on a fertilizer bag indicate different things:**

**N** Nitrogen promotes leaf growth.

**P** Phosphorus is for the formation of roots, and for flower, seed, and fruit growth.

**K** Potassium increases overall cell health. It helps plants to withstand stress from drought or cold.

There are many different formulas, but as a general rule, fertilizers with a low first number like 6-12-12 or 5-10-15 are good for new plants, and 10-10-10 is a good all-around fertilizer. Remember, whether you use organic or synthetic (manmade) types, the plants don't know the difference.

Gardening in the South is a challenge when we consider our hot summers, long periods of drought, and high humidity, but it is also rewarding with roses that bloom as late as November and daffodils that begin to bloom in February. Whether you garden in pots or have a small front yard, I hope this book will inform and inspire you to grow all types of plants. And remember, it's all about the right plant for the right place.

Annuals like Red Salvia and Marigolds have been grown for generations but gardeners today have a wealth of choices when it comes to choosing annuals that thrive in sun. Whether you limit your gardening to a few large pots for seasonal color, a window box, or an elaborate mix of annuals and perennials in your mixed border, annuals can help carry the garden through the seasons. Another way to use annuals is to combine them with your vegetables and herbs.

Many have been developed for their long season of bloom and the ability to thrive in our extreme heat and humidity. Some will also tolerate periods of drought. Star Flower acts like a butterfly magnet with its bright flowers. With others like the Sun Coleus, it's their foliage that provides color from spring until frost. Selections of Lantana and the Narrow Leaf Zinnia produce flowers all summer and into fall and they don't require deadheading. Pansies and Violas brighten the winter landscape and can be grown in combination with edibles like Lettuce or Swiss Chard or on their own.

No matter which annuals you grow, all will benefit from a well-drained soil and being watered and fertilized on a regular basis.

# SUMMER SNAPDRAGON
*Angelonia angustifolia*

With flowers that resemble snapdragons or miniature orchids, *Angelonia* thrives in the heat of summer. It attracts bees and butterflies, offers months of color, and requires no deadheading or special care. Great in containers with other annuals or in the garden, the Serena series produces compact plants that are well branched with spikes of bright flowers (1 inch across) in lavender, pink, purple, or white.

**Size:** 12 to 20 inches tall and up to 2 feet across depending on the cultivar.

**Conditions:** Full sun is ideal, although plants will still bloom in part shade. Plant this annual in a moist, well-drained soil. If it gets leggy, prune it back and it will reflower in several weeks.

**Uses and Companions:** Plant Summer Snapdragon in containers or in the border with Coleus, Petunias, Trailing Verbena, or ornamental grasses.

# TRAILING PETUNIA 'MILLION BELLS™'
*Calibrachoa × hybrida*

This charming annual looks like a miniature Petunia and quickly forms a compact mound about 4 to 6 inches tall or taller of 1-inch trumpet-shaped blooms. It flowers all summer until frost, loves heat and humidity, and doesn't need deadheading. Use it as a groundcover for the flower border, or let it spill out of hanging baskets. The Million Bells™ series includes 'Cherry Red', 'Yellow', 'Cherry Pink', and one of my favorites, 'Terra Cotta', whose flowers are a blend of yellow, bronze, apricot, and brick red.

**Size:** In the South, plants can grow 6 to 12 inches high with a spread of 24 to 36 inches.

**Conditions:** *Calibrachoa* needs full sun or part shade, though in part shade they will not flower as heavily. Plant them in average, well-drained soil.

**Uses and Companions:** Plant Trailing Petunia in the ground or in containers with ornamental grasses and Coleus. It also makes a dramatic hanging basket. Use it as an edger in the perennial border for long seasonal color.

# SPIDER FLOWER
*Cleome hassleriana*

This old-fashioned flower still holds up in the garden today. Both common names, Spider flower and Needle and Thread, refer to the way the flowers look. The curious fragrant blooms come in shades of white, pink, purple, and red. Tall and airy in the garden, it is a "see-through plant." Once established, it will tolerate some drought. Popular selections include 'Pink Queen', 'Purple Queen', 'Helen Campbell' with white flowers, and 'Linde Armstrong', a dwarf selection to 18 inches that offers flowers that have white tips in bud opening to pink.

**Size:** Plants will grow 3 to 6 feet tall and 1 to 2 feet wide. Dwarf forms only reach 18 inches at maturity.

**Conditions:** Spider Flowers perform best in full sun in well-drained soil that is not too rich. They rarely need staking. If they get too leggy, cut them back by one-third and they should continue to bloom. Many self-sow but seedlings are easy to pull out.

**Uses and Companions:** Plant Spider Flower in the middle or back of the border against an evergreen background or fence. Combine it with plants like blue Salvia and Lambs Ear. For containers, use dwarf selections.

# ECHEVERIA 'HENS AND CHICKS'
*Echeveria* spp. and cultivars

Native to Mexico, this group of succulents is both heat and drought tolerant. They offer striking architectural forms and handsome foliage ranging from silver to blue-green to almost black. They also produce spikes of colorful (often orange-pink) flowers. Separating offsets is an easy way to propagate many of the different varieties. A popular selection known as 'Topsy Turvy' has silver-green leaves that become twisted and curved as they age.

**Size:** Many are compact in habit, growing 8 to 12 inches wide and 6 to 8 inches tall.

**Conditions:** Plant these succulents in full sun or light shade in well-drained soil. One-half inch of gravel on top of the soil makes good mulch.

**Uses and Companions:** Echeverias make great container plants on their own or in combination with other succulents.

# GLOBE AMARANTH

*Gomphrena globosa*

Tough and durable, Globe Amaranth is beautiful and easy to grow in a range of situations. The fresh flowers feel like dried flowers and, in fact, Globe Amaranth is popular as a dried flower. Numerous forms include the low-growing Buddy series or the Gnome series (including 'Gnome Purple' and 'Gnome White'). A popular selection both for the garden and as a cut flower is 'Strawberry Fields' with bright strawberry-colored blooms that hold up in the heat and humidity. All have a long season of bloom and a high tolerance to drought.

**Size:** The size of the plants ranges from 12 to 24 inches with a spread of 10 to 12 inches.

**Conditions:** Plant in full sun in well-drained soil.

**Uses and Companions:** Plant Globe Amaranth at the front or middle of the border, in pots with other annuals, or in the vegetable garden. It makes a great dried flower for "tussie mussies" or other arrangements. Combine it with Salvias and Cosmos.

# LANTANA 'MISS HUFF'

*Lantana camara*

Once established, this vigorous cultivar tolerates extreme heat and drought. A nonstop bloomer from early summer until frost, it brightens the landscape with colorful orange, yellow, and pink flowers. A tender perennial in some parts of the South, plants can reach 3 to 5 feet high and 10 feet wide in one season. There are many additional selections of lantana of various sizes including 'Lemon Drop' with light yellow flowers, 'Lola' with lemon yellow flowers, 'Fran's Red' with red and yellow flowers, 'New Gold' with gold flowers, and 'Trailing Lavender, with lavender blooms.

**Size:** Lantana is 1 to 6 feet by 1 to 10 feet, depending on the selection.

**Conditions:** Plant Lantana in full sun in well-drained soil. It will tolerate a range of soil types including those with a high percentage of clay or sand.

**Uses and Companions:** Add Lantana to the perennial garden for summer color or use it in containers with other annuals like Coleus, *Angelonia*, Sweet Potato Vine, and Yucca.

# STAR FLOWER
*Pentas lanceolata*

Star Flower blooms nonstop for months, attracts butterflies and hummingbirds, and tolerates drought—all reasons to grow this durable annual. It also makes a good cut flower. Over the years many selections have been bred for vigor and flower color in shades of white, red, and pink. Two selections noted for these characteristics are 'Butterfly Pink' and 'Butterfly Red'. For a vigorous dwarf selection that grows 8 to 10 inches tall, try 'New Look Pink' or 'New Look Red'.

**Size:** Star Flower grows 12 to 36 inches by 24 inches.

**Conditions:** Plant Star Flower in full sun in average garden soil that is well drained. A midsummer haircut may increase flower production.

**Uses and Companions:** *Pentas* are useful for adding summer color to the flower border or for combining in pots with other annuals. Combine them with foliage annuals like silver Plectranthus, Salvias, or ornamental grasses.

# SUN COLEUS
*Solenostemon scutellarioides*

Coleus have long been grown for their colorful foliage and ability to grow in our heat and humidity. Recently, a whole new group that thrives in the sun has been tested and evaluated. They vary in size, the foliage comes in a range of color, and they are easy to grow and maintain. Make sure you are getting a Coleus variety for sun before purchase. A few cultivars include 'Red Ruffles' with bright red-and-burgundy foliage, 'Alabama Sunset' with brick-red leaves with a yellow edge, and 'Pineapple' with leaves that are bright lime-gold with burgundy.

**Size:** Depending on the cultivar, Coleus can range in size from under 12 inches to over 3 feet tall and wide.

**Conditions:** Plant them in full sun or part shade in soil that is moist but well drained. Pinching leggy stems will result in bushier, fuller plants.

**Uses and Companions:** Use Coleus as bedding plants, planting masses of one type or combining different varieties. Combine them in large containers with tropicals like Elephant Ears or other annuals such as *Angelonia*.

# PANSIES

*Viola × wittrockiana*

What is winter in the South without cheerful Pansies? The flowers come in many shades and combinations including white, yellow, burgundy, purple, orange, and blue. Typically they are planted in autumn and will bloom until spring or later depending on air and soil temperatures. They are mostly cold hardy and, once established, will revive after freezing temperatures and continue to bloom. There are many selections and forms from large and ruffled to small and delicate. Popular cultivars include 'Antique Shades' and 'Majestic Giants Hybrid'. The Johnny Jump Ups, *Viola tricolor*, are also charming.

**Size:** Pansies grow 8 to 12 inches depending on the cultivar. Plants should be spaced 6 inches apart.

**Conditions:** Plant Pansies in full sun or light shade in a moist, well-drained soil. For best blooms fertilize once a week.

**Uses and Companions:** Use Pansies as bedding plants or in containers. Combine them with Snapdragons, Parsley, Mustard Greens, and herbs like Golden Thyme.

# NARROW LEAF ZINNIA

*Zinnia linearis*

This colorful annual has all the charm of large Zinnia flowers and none of the problems. Even without deadheading, this free-flowering annual produces masses of miniature, daisylike golden-orange flowers from late spring until frost. Bees, butterflies, and humans alike are drawn to its colorful blooms. The Profusion series grows about 8 inches tall by 20 inches wide and has no pest or disease problems. 'Profusion Cherry' has cherry-rose flowers and Profusion 'Deep Apricot' offers peach flowers. 'Crystal White' produces 2½-inch, bright-white daisylike blooms.

**Size:** 6 to 12 inches by 24 inches wide.

**Conditions:** Plant Narrow Leaf Zinnia in full sun in well-drained soil. Deadheading will increase flower production but they will still produce lots of blooms even if you leave them alone.

**Uses and Companions:** The Narrow Leaf Zinnia is perfect for adding to the border where it can weave in and around perennials that may have already bloomed. Intense bright colors and delicate foliage add up to a useful plant for the summer garden. You can also grow it in containers with other annuals.

TEXAS FIREBUSH

# MORE ANNUALS
## FOR SUN

*Antirrhinum majus* — Snapdragons

*Brugmansia* spp. and cultivars — Angel's Trumpet

*Cosmos bipinnatus, C. sulphureus* — Mexican Aster and Yellow Cosmos

*Cuphea llavea* — Bat Face

*Hamelia patens* — Texas Firebush

*Petunia* — Wave Series

*Ruellia brittoniana* — Texas Petunia

*Salvia coccinea* — Scarlet Sage

*Scaevola aemula* — Fan Flower

*Tithonia rotundifolia* — Mexican Sunflower

# WINGED WONDERS

There is something magical about butterflies that captivates gardeners of all ages. Maybe it's their colorful wings or the way they flit from flower to flower in search of nectar. What's really amazing is that in just a few short weeks, they progress from egg, to caterpillar, to chrysalis, to beautiful winged creature.

Did you know that after bees and wasps, butterflies are important pollinators of native trees and flowers? By pollinating these plants they help perpetuate the species of certain wildflowers. Butterflies are threatened when their habitat is destroyed because of herbicide use, often associated with farming and roadside spraying. When host plants (the food for caterpillars) are destroyed, we don't get to enjoy these winged wonders.

BUTTERFLY ON SUMMER PHLOX

To find out more about butterflies, visit a butterfly house like the one at Callaway Gardens in Pine Mountain, Georgia, which is home to the Day Butterfly Center. With 4½ acres, the center includes a conservatory containing nearly 1,000 tropical butterflies of fifty different species. Visitors stroll through this rainforest environment surrounded by butterflies, tropical plants, and birds. Highlights include a 12-foot-high cascading waterfall. A unique climate control system maintains a constant temperature of 84°F. and a humidity level of 60 to 80 percent, which creates a welcoming habitat for these beautiful creatures. The area outside the center is planted with flowers and shrubs that attract native butterflies such as monarchs and swallowtails.

ZINNIA
'WHITE STAR'

Annuals offer an easy, inexpensive way to add seasonal color to your landscape, and there are dozens of choices for the shade garden. Grow them in decorative pots, hanging baskets, the mixed border, or a flower garden. Depending on your interest and comfort level, they can be purchased in six-packs or started from seed. During the heat of summer when many perennial blooms have come and gone, annuals help fill the gap both with their flowers and foliage. Some are big and bold like the many varieties of *Alternanthera*, while others are diminutive and dainty like Sweet Alyssum, *Lobularia maritima*. If they are happy, certain types like Impatiens will reseed year after year, providing a constant source of plants for your garden. Others, like Begonias, can be potted up and overwintered indoors.

Other than flowers, colorful foliage can brighten the border, woodland edge, or container. Combining shade annuals like colorful Coleus or Clown Flower with perennials such as Ferns and Hostas is another way to create dynamic garden scenes.

There are annuals for fall and winter too, like *Violas*, which include the colorful Pansies. With careful planning you can have annuals for every season.

# AGERATUM OR FLOSS FLOWER
*Ageratum houstonianum*

Old-fashioned and easy to grow, clusters of small soft-blue to violet-purple flowers make this annual a winner for the summer garden. Another reason to consider this plant is that it will tolerate periods of drought and deer don't seem to bother it. Historically it has been grown as an edging plant but there are selections like 'Blue Horizon' that grow up to 30 inches tall. In the South, Floss Flower prefers afternoon shade.

**Size:** Depending on the cultivar, Floss Flower can grow 6 to 30 inches tall. *Ageratum houstonianum* 'Artist Blue' and 'Artist Purple' grow 8 to 12 inches tall and should be spaced 10 to 12 inches apart.

**Conditions:** Plant *Ageratum* where it receives morning sun and afternoon shade. Plants bloom continuously and the Artist series don't require deadheading. Fertilize monthly with a liquid fertilizer.

**Uses and Companions:** Use *Ageratum* as an edger, in the middle of the border, in containers, or in window boxes. Combine it with other old-fashioned plants like Hollyhocks or Spider Flowers.

# CALICO PLANT
*Alternanthera* spp. and cultivars

Colorful foliage is a mainstay for summer in southern gardens where heat and humidity take its toll. Although Calico Plant has been around a long time, over the years the hybrids have gotten more colorful and many of the selections, like the upright vigorous grower 'Party Time', with pink, green, and white leaves, are better adapted for shade where they light up the garden. Other colorful cultivars include 'Grenadine', whose foliage is burgundy with hot-pink veins, and 'Crème de Menthe', with leaves that are variegated cream and green.

**Size:** Calico Plant ranges from short bedding types, reaching only 9 inches tall like the old-fashioned Joseph's Coat, *Alternanthera ficoidea*, to the large, purple-leaved 'Wave Hill', *Althernanthera dentata* var. *rubiginosa*, which can grow 3 to 4 feet tall.

**Conditions:** All plants in the genus *Alternanthera* will grow happily with morning sun and afternoon shade. The series that includes 'Party Time', 'Crème de Menthe', 'Grenadine', and 'Cognac' performs beautifully in shade.

**Uses and Companions:** Combine these colorful plants with Impatiens, Caladiums, and Angel Wing Begonias. Use them as bedding plants, in containers, or hanging baskets.

# BEGONIA: ANGEL WING, REX, BOLIVIAN

*Begonia* spp. and cultivars

Fantastic foliage and bright flowers all summer are reason to grow annual Begonias. Favorites include Angel Wing types, *Begonia coccinea*, with distinct angel wing–shaped leaves, of different shades of green with specks of silver; Rex Begonias, *Begonia rex-cultorum*, with dramatic dark leaves marked with silver; and the species, *Begonia boliviensis*, with 2-inch-long, bright orange-red bell-shaped flowers and shiny, angel wing–type leaves. These examples of the wide range of Begonias add drama and color to the summer shade garden. They also make colorful houseplants if you decide to overwinter them indoors.

**Size:** Angel Wing–type Begonias like *Begonia coccinea* and its hybrids range from 12 inches to 3 feet tall. *Begonia* x *hybrida* 'Dragon Wings' grows 2 to 3 feet tall with 5-inch-long drooping leaves. The species *B. boliviensis* grows 12 inches tall and forms 2-foot-wide clumps with long, arching canes.

**Conditions:** These grow in full shade but they tolerate full sun provided they get plenty of water. Plant in well-drained soil. Fertilize regularly and pinch back to keep them full and blooming.

**Uses and Companions:** Use them in hanging baskets, containers, window boxes, or in the border with other annuals and perennials.

# WAX BEGONIA

*Begonia semperflorens-cultorum*

These adaptable and colorful annuals will grow in full shade, part shade, or full sun provided they get enough water. They also tolerate heat and humidity better than many annuals. Flowering from late spring until frost, they offer nonstop color and require very little maintenance. The Cocktail series has bronze foliage and grows 6 to 8 inches tall with flowers of pink, rose pink, red, or white. The 'Pizzazz' series are heavy bloomers growing 8 to 10 inches tall with red, pink, or white flowers with green foliage.

**Size:** Wax Begonias range from 6 to 16 inches tall.

**Conditions:** As with most annuals, Wax Begonias perform better if they are planted in soil that is moist but well drained. Amend the soil with compost before you plant and fertilize regularly to ensure better blooms. Pinch back if they get leggy to keep plants full.

**Uses and Companions:** Use as bedding plants at the front of the border, in hanging baskets, window boxes, or in containers with other annuals. Combine them with plants like *Asparagus densiflorus* 'Sprengerii', also called the Emerald Fern, for a contrast in texture and form.

# CALADIUM OR ANGEL WINGS

*Caladium* × *hortulanum* cultivars and selections

Caladiums are tropical foliage plants from South America with colorful decorative leaves. Caladiums are tubers (an enlarged underground root for storing food) that require warm soil. Indispensable in the summer shade garden, the heart-shaped leaves are white, red, pink, or green. They may be spotted, speckled, or streaked and range in size from 12 inches to 30 inches tall. There are also dwarf selections with smaller leaves. Undemanding and easy to grow, they put on a show for months.

**Size:** Some caladiums only grow 12 inches and others grow up to 30 inches tall. 'Red Flash' is 15 to 22 inches tall, while 'Aaron' grows 15 to 20 inches tall.

**Conditions:** Caladiums require warm soil, about 65 to 70°F at a depth of 6 inches. Plant tubers one per square foot, at a depth of 2 to 3 inches, in shade or part shade. Caladiums in full sun must receive a tiny, steady drip of water. Plant in well-drained soil rich in organic matter.

**Uses and Companions:** Caladiums are great for containers or in the garden. Impatiens, Ferns, and Hostas are all happy companions for Caladiums.

# BEDDING IMPATIENS: NEW GUINEA IMPATIENS AND BUSY LIZZIE

*Impatiens walleriana* and *Impatiens hawkeri*

Impatiens can light up a garden with their brilliant flowers. Blooming from spring until frost, Impatiens and shade go together like milk and cookies. Although it is an annual, if it is happy in your garden, it will reseed and spread itself around from year to year. The wide range of colors, from the purest white to the most vivid pink or red, single-flowered or double-flowered types, and even some with variegated foliage, make it hard to choose a favorite. The bedding types can get quite large but there are dwarf types that work well in containers.

**Size:** Impatiens grow anywhere from 6 inches to 3 feet tall and wide. The Fiesta series only get 12 inches tall. New Guinea hybrid 'Tango' has 2½ inch blossoms and grows 2 feet tall and 18 inches wide.

**Conditions:** Plant 8 to 10 inches apart. Impatiens are happiest in moist, well-drained soil. Even the sun-tolerant New Guinea hybrids benefit from afternoon shade. Fertilize every two weeks with a liquid fertilizer.

**Uses and Companions:** Use Impatiens as bedding plants, in containers in combination with other shade-loving annuals like *Caladiums*, Ferns, or *Torenia*.

# SWEET ALYSSUM
*Lobularia maritima*

The tiny flowers pump out a fragrant honey scent that is a delight to experience on hot summer days. Although the blooms and foliage look delicate, this plant is a survivor. It makes the perfect edger for the summer border where it can weave itself in and out among other annuals and perennials.

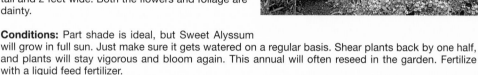

**Size:** This branching, trailing annual grows 1 foot tall and 2 feet wide. Both the flowers and foliage are dainty.

**Conditions:** Part shade is ideal, but Sweet Alyssum will grow in full sun. Just make sure it gets watered on a regular basis. Shear plants back by one half, and plants will stay vigorous and bloom again. This annual will often reseed in the garden. Fertilize with a liquid feed fertilizer.

**Uses and Companions:** Use Sweet Alyssum to edge the front of flower borders, in between stepping stones, or as a carpet under spring bulbs. Combine it with other tall annuals.

# COLEUS
*Solenostemon scutellarioides*

Fantastic colorful foliage makes the low-maintenance Coleus a must for the shade garden, providing months of color. Some of the choice selections include 'Fishnet Stockings' with striking green and burgundy variegation that grows 24 to 36 inches tall. 'Freckles' has creamy yellow foliage with orange and bronze splotches and grows 10 to 14 inches tall. 'India Frills' grows 15 inches tall and acts more like a carpet with finely edged leaves lined in lime green and splashed with pink, purple, and orange. 'Thumbelina' has a compact habit and tiny rounded leaves of green and burgundy.

**Size:** Depending on the cultivar, Coleus can range in size from under 12 inches to over 3 feet tall and wide.

**Conditions:** Plant shade-tolerant varieties in part or full shade in moist but well-drained soil. Removing the flowers will keep plants looking healthier longer. Pinching leggy stems will also result in bushier plants. Fertilize once in June, July, and August using half-strength liquid fertilizer.

**Uses and Companions:** Plant them in large decorative pots, window boxes, hanging baskets, or as bedding plants. Combine them with other Coleus, trailing plants like variegated Vinca, Impatiens, Ferns, Hostas, and tropical plants like Bananas or Elephant Ears.

# BACOPA
*Sutera* hybrids and cultivars

Sometimes it's the smaller, more delicate blossoms that serve myriad functions in the garden. For hanging baskets and containers, *Bacopa* is perfect. Vigorous and easy to grow, this trailing plant combines well with bigger plants. Its lacy green foliage and flowers in shades of white, lavender, and blue provide an ideal filler and quickly spill over the sides of their container. Even without deadheading, *Bacopa* pumps out color all summer long. 'Snowstorm', or 'Snowflake', has white flowers and 'Glacier Blue' has blue flowers and more of an upright habit growing 6 to 10 inches tall.

**Size:** *Bacopa* has tiny foliage and blossoms; plants generally grow 4 to 8 inches tall and should be planted 8 to 10 inches apart.

**Conditions:** Plant *Bacopa* in a sunny location that receives afternoon shade in moist, well-drained soil. *Bacopa* is not a plant that tolerates drought. You'll know plants are stressed if they drop all their flowers and buds. If this happens, they should recover in about two weeks.

**Uses and Companions:** *Bacopa* is perfect for hanging baskets, window boxes, or containers. To really show off their blooms, combine *Bacopa* with larger flowers and foliage of plants such as Petunias, Coleus, and Verbena.

# WISHBONE FLOWER OR CLOWN FLOWER
*Torenia fournieri*

Known as the Wishbone Flower for its wishbone-like fused stamens, or the Clown Flower, this annual puts on a colorful show in the garden, blooming for months from spring until frost. The Summer Wave series are vigorous hybrids that bloom nonstop with unusually large flowers. 'Summer Wave Blue' and 'Summer Wave Amethyst' will thrive in sun or shade and are not bothered by heat and humidity. If plants get too large, you can cut them back and they will continue to pump out colorful flowers. They are great for containers or as bedding plants.

**Size:** Wishbone Flower grows 8 to 10 inches tall and should be spaced 8 to 10 inches apart.

**Conditions:** Plant Wishone Flower in part shade or sun. In full sun situations make sure it receives ample moisture. Shear plants back if they become leggy or too large. Fertilize with a liquid fertilizer once a month.

**Uses and Companions:** Wishbone Flower makes an excellent choice for a container plant or hanging basket. It also works well in the front of the perennial border. Combine it with Ferns like *Asparagus* 'Sprengeri' or variegated trailing Vinca.

PEACOCK GINGER

# MORE ANNUALS
## FOR SHADE

*Asparagus densiflorus* 'Sprengeri'—*Asparagus Fern*

*Coleus*—*many different cultivars*

*Dianella tasmanica* 'Variegata'—*Variegated Flax Lily*

*Hedychium gardnerianum* 'Kahili'—*Kahili Ginger*

*Hypoestes phyllostachya*—*Polka Dot Plant*

*Kaempferia pulchra* 'Mansoni'—*Peacock Ginger*

*Nephrolepis exalta* 'Rita's Gold'—*Rita's Gold Boston Fern*

*Ruellia brittoniana*—*Mexican Petunia*

*Setcreasea pallida* 'Purple Heart'—*Purple Heart*

*Strobilanthes dyerianus*—*Persian Shield*

# VARIETY IS THE SPICE OF GARDENING

What is variegation? In plants, it is when the normal green portion of the plant leaf is replaced by colors of white, cream, yellow, or on occasion, other colors. The variegation can occur in the center or along the edges of a leaf.

There are different causes for variegation, ranging from virus infections to visual effects due to reflected light. Whatever the cause, the result is the same—interesting and colorful foliage in shrubs, trees, and herbaceous plants that light up the landscape. Use variegated plants as focal points, or plant a group of them together against a backdrop of evergreens. Consider these variegated selections for your garden.

WILD GINGER

## Herbaceous Plants
*Asarum splendens*—Showy Wild Ginger
*Carex oshimensis* 'Evergold'—Variegated Sedge
*Yucca filamentosa* 'Color Guard'—Color Guard Yucca

## Shrubs
*Acanthopanax sieboldianus* 'Variegatus'—Variegated Five-leaf Aralia
*Aralia elata* 'Variegata'—Japanese Angelica
*Cercis canadensis* 'Floating Clouds'—Variegated Redbud
*Chameacyparis pisifera* 'Filifera Aurea'—Golden Threadleaf
*Osmanthus heterophyllus* 'Goshiki'—Variegated Holly-Leaved Osmanthus

## Trees
*Cornus controversa* 'Variegata'—Giant Dogwood
*Cornus florida* 'Daybreak'—Variegated native Dogwood
*Cornus kousa* 'Wolf Eyes'—Variegated Kousa Dogwood
*Robinia pseudoacacia* 'Frisia'—Frisia Black Locust

## Vines
*Kadsura japonica* 'Variegata'—Variegated Kadsura
*Parthenocissus tricuspidata* 'Fenway Park'—Golden Leaf Boston Ivy
*Schizophragma hydrangeoides* 'Moonlight'—Japanese Hydrangea Vine

HYPOESTES PHYLLOSTACHYA
POLKA DOT PLANT

# FLOWERING PERENNIALS

Although they are not permanent fixtures in the garden, hardy perennials reward gardeners year after year. Many are happiest with full sun or part sun exposure. Some, like the Japanese Roof Iris, are evergreen but many die back to the ground at the end of the growing season and put up new leaves and flowers every year. There are types that bloom in spring, summer, and fall. Some provide interesting textures and others display colorful foliage in autumn too. Once established, many can be easily divided and shared with friends.

Before you plant, make sure the soil is enriched with organic matter and well drained. Plan ahead if you intend to grow perennials that require staking, or site them in the garden so that other plants will support them as they grow up.

Perennials provide perfect companions for other perennials, shrubs, and trees. Some, like Garden Phlox, are grown for their big beautiful blooms, while others, like Japanese Aster, produce many small flowers over a long period of time.

By growing a range of different perennials and combinations of perennials in your garden, you can have attractive flowers or foliage throughout the year—the gardener's dream!

# ARKANSAS BLUESTAR
*Amsonia hubrichtii*

From spring until frost, Arkansas Bluestar offers great texture, grace, and beauty to the perennial garden with its threadlike leaves. In spring the pale blue, almost white, flowers are a bonus. The real show, though, is in autumn, when the foliage turns golden yellow and then brown. Once established, this native is drought friendly and requires a minimum of care. In just a few years, you will have a big clump that you can divide and share with friends.

**Size:** Arkansas Bluestar grows 3 feet tall by 3 feet wide.

**Conditions:** Full sun and moist, well-drained soil is ideal but Arkansas Bluestar will tolerate drier soils as well.

**Zones:** 6 to 9

**Uses and Companions:** Combine *Amsonia hubrichtii* with other perennials like *Geranium* 'Rozanne' and *Baptisia*. I have enjoyed the fall colors when it is paired with purple Asters and *Callicarpa americana,* American Beautyberry, which also has purple fruits.

# ASTER
*Aster oblongifolius* 'Raydon's Favorite' and *Aster tataricus*

Asters are a large group of plants, making it hard to choose the best, but 'Raydon's Favorite' and *Aster tataricus* are two reliable performers in Southern gardens. 'Raydon's Favorite', a selection of one of our native asters, has purple daisy-type flowers with yellow centers and blooms in fall, usually September to October. *Aster tataricus* is a tall species that doesn't require staking and its flowers are most welcome, often appearing as late as November.

**Size:** 'Raydon's Favorite' grows 2 to 3 feet tall and 2 to 3 feet wide. *Aster tataricus* grows 3 to 6 feet tall and has large basal leaves.

**Conditions:** These like full sun and well-drained soil. Propagate by division every few years in early spring or fall; this will help to keep plants vigorous.

**Zones:** 3 to 8

**Uses and Companions:** Plant Asters in the perennial border. Combine them with Goldenrod, Salvias, and other Asters.

# FALSE INDIGO

*Baptisia* 'Carolina Moonlight'
and 'Purple Smoke'

There are a number of choice *Baptisia* species and hybrids. Two of my favorites are 'Carolina Moonlight' and 'Purple Smoke', selections bred by Rob Gardner of North Carolina Botanical Garden in Chapel Hill, from crosses of *B. sphaerocarpa* and *B. alba*. Both have handsome blue-green foliage and give the effect of a small shrub in the garden. 'Carolina Moonlight' has light yellow flowers and 'Purple Smoke' has smoky-purple flowers that last for weeks. The flowers resemble Lupines but are much better suited for our climate.

**Size:** Both 'Carolina Moonlight' and 'Purple Smoke' grow 3 to 4 feet tall and wide.

**Conditions:** They like full sun and well-drained soil. Propagate by division in early spring or fall.

**Zones:** 4 to 9

**Uses and Companions:** Plant *Baptisia* in the perennial border. Underplant with groundcovers like *Veronica* 'Georgia Blue' or annuals like *Euphorbia* 'Diamond Frost'. Pair *Baptisia* 'Purple Smoke' with Peonies.

# CONEFLOWERS

*Echinacea* spp. and cultivars

Purple Coneflower, *Echinacea purpurea*, is a native that has been popular in gardens for years. The 3- to 4-inch-diameter flowers have a center cone that resembles a miniature hedgehog, surrounded by purple petals. Blooming in summer, this tough perennial attracts butterflies and hummingbirds alike and tolerates our heat and humidity. There are numerous selections available, including those with red, orange, or yellow flowers.

**Size:** Coneflowers range 2 to 4 feet tall depending on the cultivar. Plants form clumps that are 2 feet wide.

**Conditions:** Full sun and well-drained soil is best. Don't give these perennials any extra fertilizer. Deadheading may encourage more blooms.

**Zones:** 3 to 8

**Uses and Companions:** Coneflower is great in the perennial border with other summer bloomers like Garden Phlox, Salvias, and Daylilies. You can mix and match the different selections like *Echinacea* 'Sundown' with orange-red flowers and *E. purpurea* with purple flowers.

# GUARA OR WHIRLING BUTTERFLIES
*Gaura lindheimeri*

These are tough plants for tough times. A native of Texas, *Gaura* is a good performer despite heat and humidity. It also tolerates poor soils, is drought friendly, and will thrive in places like the strip between the sidewalk and the street where few other plants dare to grow. The white flowers, 1 to 3 inches long, are tinged rose and rise well above the foliage, reminding some of butterflies, hence its other common name of Whirling Butterflies.

**Size:** *Guara* ranges 3 to 4 feet tall by 3 feet wide.

**Conditions:** Full sun and well-drained soil are best but *Gaura* can take some abuse. Cut back the long, wiry flower stems after these finish blooming to encourage another flush of flowers. Divide plants in spring if they become too large.

**Zones:** 5 to 8

**Uses and Companions:** I planted this perennial on the strip between the sidewalk and the street where plants need to be tough. Combine it with other drought-friendly plants like Yuccas, ornamental grasses, and Salvias. Grow it in the border or rock garden.

# CRANESBILL
*Geranium 'Rozanne'*

I first grew this plant after admiring it in a garden in Asheville, North Carolina. The gardener dug it up for me in June and I brought it home to Georgia. It continued to bloom until November. Several years later it is still a favorite of mine for its long bloom period, colorful blue-violet flowers, and trailing habit. It also tolerates heat and humidity and responds well to being cut back if it starts to loose vigor.

**Size:** Cranesbill grows 18 inches tall by 24 inches wide, forming a mound with trailing stems.

**Conditions:** Full sun and moist, well-drained soil are ideal. Cut back trailing stems to rejuvenate plants or encourage more blooms.

**Zones:** 5 to 8

**Uses and Companions:** This is a much better choice for Southern gardens than *Geranium* 'Johnson's Blue', which does not do well in our extreme heat and humidity. Combine it with white Japanese Roof Iris, Roses, and Lambs' Ears.

# JAPANESE ROOF IRIS
*Iris tectorum*

Vigorous and easy to grow, I like this plant both for its blooms and its foliage, which looks good all season long. The large, beardless flowers—up to 6 inches wide—are medium-blue to purple and speckled with darker purple. They appear in May to June with the 6- to 8-inch-wide foliage. The selection 'Album' is even more beautiful with its white flowers marked with yellow.

**Size:** Japanese Roof Iris is 12 to 18 inches tall by 18 inches wide.

**Conditions:** Full sun and moist, well-drained soil are ideal but this Iris will tolerate drier soils as well.

**Zones:** 4 to 8

**Uses and Companions:** Plant this perennial at the front or middle of the border. Combine it with other Irises, Hardy Geraniums, and Japanese Asters.

# JAPANESE ASTER
*Kalimeris pinnatifida*

Japanese Aster provides delicate foliage and double, white daisylike flowers for months. The best part is it requires no special attention or fuss and makes a great filler in the garden. If it gets leggy you can cut it back, but it will still bloom nonstop without deadheading. Use it throughout the flower border as a transition between different colors. It makes a good cut flower for arrangements too.

**Size:** Japanese Aster grows 1 to 2 feet tall and 2 feet wide.

**Conditions:** Full sun and well-drained soil are best but Japanese Aster will also grow and bloom in part shade. Divide plants in early spring or fall if the clumps become large, floppy, and start to split open.

**Zones:** 4 to 8

**Uses and Companions:** Plant Japanese Aster in the middle or back of the flower border. Combine it with Garden Phlox, *Crocosmia*, Blanket Flower, Iris, and Hardy Geraniums.

# WHITE GARDEN PHLOX
*Phlox paniculata 'David'*

Garden Phlox have long been a mainstay of the summer border but many suffer in our heat and humidity and are prone to powdery mildew. This is not the case with the selection called 'David', which is a strong grower and bloomer. The white flowers brighten the garden for weeks and add height, too. It's worth noting that if you remove the spent flowerheads, you may get a second flush of blooms.

**Size:** Phlox 'David' grows 2 to 3 feet tall and 2 feet wide.

**Conditions:** Full sun, well-drained soil, and good air circulation are the keys to success with this plant. Avoid overhead watering if possible; this will help reduce the chances of plants getting mildew. Divide plants in early spring if needed.

**Zones:** 4 to 8

**Uses and Companions:** White flowers in the garden are soothing, especially during the heat of summer. Garden Phlox attract butterflies and add height and color. Combine Phlox with Lilies, Hardy Geraniums, Lavender, and Iris.

# BLACK-EYED SUSAN OR ORANGE CONEFLOWER
*Rudbeckia fulgida 'Goldsturm'*

From late summer into fall, *Rudbeckia* 'Goldsturm' brightens the garden with its 3- to 4-inch-wide, rich yellow-gold, daisylike flowers. These are set off by a nearly black center cone and dark green foliage. Heat and humidity don't seem to bother this coneflower. A fast grower, it quickly forms large clumps that can be easily divided in spring or fall and shared with friends. This plant is also known as the Orange Coneflower. Black-eyed Susan also makes a good cut flower.

**Size:** Black-Eyed Susan grows 1½ feet to 2½ feet tall by 2 feet wide.

**Conditions:** Full sun and moist, well-drained soil are best but once established, this perennial is drought friendly and will survive periods with limited amounts of water.

**Zones:** 3 to 8

**Uses and Companions:** Plant 'Goldsturm' in the perennial border with other red and yellow flowers for late summer and fall color. Combine it with Asters, Salvias, Daylilies, and Lambs' Ears.

VERONICA 'GEORGIA BLUE'

# MORE
# PERENNIALS
## FOR SUN

*Asclepias tuberosa*—Butterfly Weed

*Ceratostigma plumbaginoides*—Hardy Plumbago

*Coreopsis auriculata* cultivars—Mouse Ear Coreopsis

*Hemerocallis* spp. and cultivars—Daylilies

*Leucanthemum × superbum* 'Becky'—Becky Shasta Daisy

*Rudbeckia triloba*—Three-Lobed Coneflower

*Salvia guaranitica*—Blue Anise Sage

*Sedum spectabile* 'Autumn Joy'—Showy Stonecrop

*Stokesia laevis* cultivars—Stokes Aster

*Veronica* 'Georgia Blue'—Creeping Speedwell

# MORE ABOUT BUTTERFLIES

## HOW TO ATTRACT BUTTERFLIES

Attract butterflies to your garden by planting a combination of host and nectar plants.

### Host Plants (larval food plant)

Butterfly caterpillars are very specific about the plants they eat. Some common butterflies and their preferred host plants include:

**Monarchs**—Milkweeds, *Asclepias* species
**Zebra Swallowtails**—Paw Paw, *Asimina triloba*
**Tiger Swallowtails**—Tulip Tree, *Liriodendron tulipifera*, or Sweetbay Magnolia, *Magnolia virginiana*
**Spicebush Swallowtails**—Spicebush, *Lindera benzoin*, and Sassafras, *Sassafras albidum*

### Nectar Plants

Visit your local nursery and observe butterflies in your area. Not all bright flowers provide nectar but here are some popular butterfly plants. Plant these with your host plants. Butterflies also need a place to sun and a source of water. Mud puddles and rocks will do the trick.

MEXICAN SUNFLOWER

**Butterfly Weed**—*Asclepias tuberosa*
**Butterfly Bush**—*Buddleia davidii*
**Buttonbush**—*Cephalanthus occidentalis*
**Coneflower**—*Echinacea* species
**Hardy Ageratum**—*Eupatoriuum coelestinum*
**Lantana**—*Lantana camara*
**Passion Flower**—*Passiflora* species
**Starflower**—*Pentas lanceolata*
**Hummingbird Sage**—*Salvia guaranitica*
**Mexican Sunflower**—*Tithonia rotundifolia*

HEMEROCALLIS SPP.

Shade can be a blessing and a curse. Beginning gardeners may find it challenging to create a colorful and interesting garden in the shade. A shady woodland is a welcome relief, especially during hot summer months, but a dense Pine forest can limit your options when it comes to growing a diversity of plants. While it's true that many perennials need lots of direct sun to ensure good blooms, there are those that offer striking foliage and flowers in the shade.

The first step is to determine how much shade you have. Does the garden receive any direct sun; if so, how many hours? As a general rule, blooming plants require at least several (4 to 6) hours of sunlight to bloom, even if it is filtered through deciduous trees. Few if any perennials will grow in darkness. For many plants, morning sun and afternoon shade is ideal, but this is not always the case. If you have large, mature trees, thinning out some of the branches is one way to allow more light into your garden. (Consult with an arborist before you begin pruning any trees.)

Combine your shade perennials with colorful shade annuals in containers, the woodland, and in hanging baskets.

# CAST IRON PLANT
*Aspidistra elatior*

Some old-fashioned favorites remain popular for good reason. This tough guy has been a staple of Southern gardens for years. Bold, wide, and dark evergreen leaves provide a strong vertical accent. Cast Iron Plant will grow in deep shade and, once established, is drought tolerant. Variegated selections include *Aspidistra elatior* 'Asahi' with leaves that are 20 inches long and 5 inches wide, the upper third turning white as the season progresses, and *Aspidistra elatior* 'Variegata' with vertical white bands of varying widths.

**Size:** Cast Iron Plant grows 2 to 3 feet by 1 to 2 feet, with upright vertical leaves.

**Conditions:** It prefers shade to heavy shade.

**Zones:** 7 to 10

**Uses and Companions:** Use Cast Iron Plant in woodland gardens, for foundation plants, or as cut leaves for flower arrangements. Planted under windows next to the house, it makes an effective foundation plant especially in combination with Ferns or Hellebores. Combine them with Hostas and Boxwoods.

# HARDY BEGONIA
*Begonia grandis* subsp. *evansiana*

Hardy Begonia is a great plant for its late summer and fall interest in the garden. Its waxy leaves look fresh all season, especially when the red undersides are backlit from the late afternoon sun. Keep in mind, though, that it will disappear at the end of the growing season and return in the spring. Airy sprays of pink flowers rise above the foliage and bloom for weeks. The seed capsules dry on the plant, adding winter interest. There is also a white-flowered form.

**Size:** Hardy Begonia grows 1 to 2 feet by 2 to 3 feet, with an upright and spreading habit.

**Conditions:** This plant prefers light to heavy shade. Spreading easily by bulblets that fall from the stem, it makes a good mass planting in no time.

**Zones:** 6 to 9

**Uses and Companions:** Plant this in the woodland with other perennials like woodland Asters or under shrubs. Combine Hardy Begonia with Hostas, Variegated Solomon's Seal, Autumn Fern, Christmas Fern, and native Azaleas.

# FRINGED BLEEDING HEART

*Dicentra eximia* 'Luxuriant'

Fringed Bleeding Heart is an elegant native that is more heat and sun tolerant than the species Giant Bleeding Heart, *Dicentra spectabilis,* which disappears after it blooms, leaving a gap in the garden. Longer blooming and more persistent in the garden, 'Luxuriant' has lacy blue-green foliage and deep pink, heart-shaped blooms that rise above the foliage from spring into fall. It forms a tidy clump about 1 to 1½ foot high.

**Size:** Fringed Bleeding Heart will grow to 1 to 1½ feet by 1 foot.

**Conditions:** This plant likes an exposure of part sun to full shade. Although it will tolerate more sun than other Bleeding Hearts, a shady site with moist, well-drained soil is best.

**Zones:** 3 to 9

**Uses and Companions:** Combine Fringed Bleeding Heart with Rhododendrons, Azaleas, Hostas, Ferns, Giant Bleeding Heart, *Epimedium*, and Hellebores.

# EPIMEDIUM OR BARRENWORT

*Epimedium* spp. and cultivars

*Epimedium* is a tough perennial that, once established, will grow in dry shade even under Maples where competition with tree roots for moisture is an issue. The handsome heart-shaped leaves start out soft and green but turn leathery as they mature. Depending on the variety, some have evergreen or persistent leaves that display tinges of red and bronze in the autumn. Spikes of tiny, Columbine-like flowers rise above clumps of foliage in early spring and come in shades of yellow, white, orange, red, and lavender.

**Size:** *Epimedium* grows 6 to 20 inches tall by 2 feet wide.

**Conditions:** *Epimedium* and Barrenwort prefer part to deep shade, and they tolerate both well-drained and dry soils, once established. A site that receives morning sun and afternoon shade is ideal. If the foliage looks tattered at the end of the growing season, cut it back to the ground in early spring.

**Zones:** 5 to 8

**Uses and Companions:** Plant *Epimedium* as a groundcover under trees, Rhododendrons, and other shrubs. Combine it with Ferns, Hellebores, and Hostas.

# LENTEN ROSE
*Helleborus* × *hybridus*

A gem in the winter garden, this elegant perennial offers beautiful flowers in late winter through early spring. It boasts handsome foliage year-round and adapts to a range of garden situations. The nodding blooms, which resemble single roses in shades of white, green, and maroon, sometimes speckled or splotched, occur over a period of eight to ten weeks. Both the leaves and flowers are stemless and arise directly from the rootstock.

**Size:** Lenten Rose grows 15 to 18 inches by 18 inches.

**Conditions:** Lenten Rose prefers part sun to shade, but it will tolerate full sun if the soil is moist and well drained. Cut back dying leaves in early spring to encourage a flush of new growth. Transplant seedlings in early spring.

**Zones:** 4 to 8

**Uses and Companions:** Combine Lenten Rose with early-blooming Daffodils in the deciduous woodland garden to mask the Daffodil foliage as it ripens later in the season. Ferns and Hostas also make happy companions. Use Lenten Rose as a groundcover in shady areas.

# CORALBELL OR ALUMROOT
*Heuchera* spp. and cultivars

Although Coralbells produce small sprays of delicate bell-shaped flowers, it is the colorful evergreen foliage that makes them stand out in the garden. Forming tidy mounds of foliage, they offer an alternative to Hostas. Many of the recent hybrids of both *Heuchera villosa* and *Heuchera americana* offer vigorous plants with striking, colorful foliage that exhibits heat and drought resistance. Some, like *H. americana* 'Dale's Strain', have silver and green-marbled leaves, while others, like *H*. 'Citronelle', offer bright yellow foliage or shades of peach and apricot as with 'Caramel'.

**Size:** Coralbells will grow to 15 to 18 inches by 18 to 24 inches.

**Conditions:** For the best results plant Coralbell and Alumroot in part shade in moist, well-drained soil. Morning sun and afternoon shade is ideal. Prune dead or tattered leaves as needed.

**Zones:** 3 to 8

**Uses and Companions:** Plant Coralbell in the woodland to brighten up a dark corner. Combine them with Hostas, Maidenhair Fern, Autumn Fern, Christmas Fern, or other Coralbells.

# HONEYBELLS HOSTA

*Hosta* 'Honeybells'

For striking foliage in the shade garden, Hostas are hard to beat. 'Honeybells' offers large (8 by 11 inches), oval, olive-green leaves and fragrant flowers, white flushed with lilac, in late summer. This vigorous selection is drought tolerant and multiplies to form big clumps very quickly. Even without the flowers, the foliage adds texture and color to the garden. Other Hostas recommended include *Hosta plantaginea* 'Aphrodite', 'Venus', 'Fried Bananas', 'Fried Green Tomatoes', and 'Guacamole'.

**Size:** Honeybells Hosta grows 24 to 30 inches by 48 inches.

**Conditions:** Morning sun and afternoon shade is ideal, but 'Honeybells' will tolerate more sun if it has plenty of moisture.

**Zones:** 3 to 8

**Uses and Companions:** Plant Hostas in a mass to create a focal point, or combine them with other Hostas and Ferns, like the evergreen Autumn Fern or perennials like *Phlox stolonifera*. This way, when the Hostas die back in winter, the Ferns will provide interest. Underplant Hostas with evergreen groundcovers like *Ajuga reptans* 'Catlin's Giant'.

# CARDINAL FLOWER

*Lobelia cardinalis*

Bright red flowers in August on 2- to 4-foot-tall spikes that attract hummingbirds are reason enough to grow this native beauty. Other reasons include its ability to grow in damp or well-drained soils. Cardinal Flower will thrive along streams, in bogs, or in the flower border.

**Size:** Cardinal Flower grows 2 to 4 feet by 2 feet in the garden.

**Conditions:** This plant likes part shade or shade in wet or dry soil.

**Zones:** 2 to 9

**Uses and Companions:** Plant Cardinal Flower in damp soils near streams or ponds. Combine it with Royal Fern, Japanese Iris, *Ligularia*, or shrubs like *Itea virginica* 'Henry's Garnet'. In the border, plant it with Joe-Pye Weed, Black-eyed Susan, and Salvias.

PROVEN PLANTS

This native offers handsome notched foliage ranging from green to blue-green with silver or white veining on each leaf. Evergreen, vigorous, and drought tolerant, over time it colonizes to form dense carpets in the woodland or rock garden. In early spring it produces small pinkish-white fragrant flowers that rise above the foliage. It is not subject to any serious pest or disease problems, unlike its Asian counterpart, *Pachysandra terminalis*.

## ALLEGHANY SPURGE
*Pachysandra procumbens*

**Size:** Alleghany Spurge grows 8 to 12 inches high by 12 to 15 inches wide.

**Conditions:** Plant this native in dappled to deep shade in moist, well-drained soil. Once established, it will tolerate periods of drought.

**Zones:** 5 to 9

**Uses and Companions:** Alleghany Spurge makes an effective groundcover in the woodland under shrubs. Combine it with other hardy perennials like Hardy Gingers, Dwarf Crested Iris, Ferns, and Hostas.

## VARIEGATED SOLOMON'S SEAL
*Polygonatum odoratum*
*var. pluriflorum* 'Variegatum'

An aristocrat for the shade garden, Variegated Solomon's Seal is that rare combination of elegance and toughness. Although slow to establish, it is well worth the wait. The variegated green-and-white leaves hang from arching stems and brighten the garden all summer. In autumn the leaves turn yellow before they fall off. It also produces small white flowers, but the foliage is the star here.

**Size:** Variegated Solomon's Seal grows 20 inches by 30 inches wide.

**Conditions:** Plant Variegated Solomon's Seal in shade to part sun in moist, well-drained soil. Once established, it will tolerate periods of drought.

**Zones:** 5 to 9

**Uses and Companions:** Combine Variegated Solomon's Seal with Ferns, Coralbells, Hostas, Hardy Gingers, and *Ajuga*. A mass of just this plant will brighten up any shade garden.

PROVEN PLANTS

CREEPING PHLOX

# MORE
# PERENNIALS
## FOR SHADE

*Acanthus* 'Summer Beauty'—*Bear's Breeches*

*Hexastylis shuttleworthii* 'Callaway'—*Mottled Wild Ginger*

*Asarum splendens*—*Wild Ginger*

*Euphorbia amygdaloides* var. *robbiae*—*Mrs. Robb's Bonnet*

*Helleborus foetidus*—*Bearsfoot Hellebore*

*Phlox divaricata*—*Woodland Phlox*

*Phlox stolonifera*—*Creeping Phlox*

*Rohdea japonica*—*Nippon Rohdea*

*Tiarella cordifolia*—*Foamflower*

*Tricyrtis formosana*—*Formosa Toad-Lily*

# WATER-WISE GARDENING

Many parts of the South have experienced years of drought, and this has had an impact on how people garden. Below are some tips for water-wise gardening.

1. **Put the right plant in the right place.** This will make a big difference in whether your plants thrive or perish. If a plant likes moist, well-drained soil, do not place it in sandy soil in full sun. Group plants together that have similar water requirements. Select plants that are "drought friendly." Perennials, like *Baptisia*, have a taproot, while Daylilies have thick storage roots. Plants with waxy coated leaves, such as many Sedums, are well equipped to tolerate periods of drought.

2. **Prepare the soil before planting.** Add organic matter or compost to the soil to help with water retention when planting annuals, perennials, and bulbs.

3. **Apply a layer of organic mulch.** Arrange a layer (2 to 3 inches) between and around plants, taking care to keep mulch away from the trunks of trees and shrubs, as well as the stems of perennials.

4. **Take caution with fertilizer.** Apply less fertilizer, and avoid those with too much nitrogen (N), which will cause excessive growth—meaning plants will need more water.

5. **Be conservative with pesticides.** Keeping pesticide applications to a minimum when plants are stressed by excessive heat and drought conserves water.

6. **Use rain barrels or cisterns to collect and store rainwater.** There are many Web sites containing detailed information about the practice of harvesting rain.

DAYLILY

*HELLEBORUS FOETIDUS*
BEARSFOOT HELLEBORE

# FERNS

What would the woodland be without ferns—graceful, tough, and long lived? Some are small and delicate, while others are big and bold, but all add their own unique architecture to the garden. Ferns are also the perfect complement for Hostas, Hellebores, Hardy Gingers, and other shade-loving perennials. Some, like the Christmas and Autumn Fern, are evergreen, adding their own unique color throughout the year. Others, such as the Japanese Painted Fern and Cinnamon Fern, put on a show that lasts from early spring, when the showy new leaves unfurl, until fall when the fronds turn yellow-brown before they wither away.

There are those that spread easily and act like groundcovers, carpeting the forest floor under trees and shrubs, while others form tidy clumps. While there are ferns that will thrive in hot, sunny gardens, the species recommended here are for shady environments with soils that are moist and well drained. Yet, once established, certain ferns are drought friendly and can survive periods without any supplemental water. The ferns in this book, each with its own unique character, are hardy perennials that should persist and perform in the garden for years without much special care.

# SOUTHERN MAIDENHAIR FERN
*Adiantum capillus-veneris*

The native Southern Maidenhair Fern is graceful, delicate, and easy to grow. The thin, wiry dark stems contrast nicely with the bright green, finely cut foliage, especially when the sunlight hits it. This lacy-leaved beauty adds elegance to the woodland.

**Size:** This fern grows 1½ feet tall and spreads by creeping rhizomes.

**Conditions:** Plant this fern in shade in a soil that is moist but well drained and rich in organic matter. Propagate by dividing mature clumps in early spring.

**Zones:** 5 to 8

**Uses and Companions:** Maidenhair fern is happiest in shady woodlands. For a contrast, combine it with perennials like *Helleborus* x *hybridus*, Lungwort, and the native *Pachysandra procumbens*. Plant it next to a pond alongside Cardinal Flower, *Lobelia cardinalis*. It also makes a stunning container plant.

# LADY FERN
*Athyrium filix-femina*

If you need a vigorous perennial to cover a large area quickly and provide vertical interest, Lady Fern fits the bill. The individual yellow-green fronds are delicate in appearance but the overall plant makes a bold statement.

**Size:** Lady Fern grows 3 to 4 feet tall or taller and spreads rapidly.

**Conditions:** Plant in part or full shade in moist, well-drained soil. This fern will also tolerate full sun provided it receives lots of moisture. Propagate by division of clumps in spring or early summer. Wait until spring to cut back the brown fronds.

**Zones:** 4 to 8

**Uses and Companions:** Plant Lady Fern in a woodland where you need to cover large areas. Combine it with other ferns and Hostas.

# JAPANESE PAINTED FERN 'PICTUM'

*Athyrium nipponicum* 'Pictum'

This elegant fern brightens a woodland with its striking foliage of grayish green and silvery dark maroon. Japanese Painted Fern is both hardy and easy to grow. The colorful fronds grow up to 24 inches tall and look good from spring until frost. This is a great plant for adding color and texture to the shade garden.

**Size:** Up to 24 inches tall and wide.

**Conditions:** Shade or part shade is ideal with a soil that is moist but well drained. Propagate by division of clumps in early spring or fall.

**Zones:** 3 to 8

**Uses and Companions:** Combine Japanese Painted Fern with Hostas, Hellebores, and perennials like *Acorus* or selections of *Heuchera* with purple foliage. This fern works great in the landscape or in containers with other perennials. Underplant Boxwood with Japanese Painted Fern for a contrast.

Japanese Holly Fern is a sturdy evergreen fern that looks more like a Holly than a fern with thick, shiny fronds that are held upright in a vaselike shape. As tough as it is, it is also elegant and bold, making a statement in the landscape especially when it is paired with more delicate plants.

# HOLLY FERN

*Cyrtomium falcatum*

**Size:** This fern grows 1½ to 2 feet tall and 2 feet wide or wider.

**Conditions:** Grow Japanese Holly Fern in full sun or part shade in a soil that is moist but well drained. Once it is established this fern will withstand extended periods without water.

**Zones:** 7 to 10

**Uses and Companions:** Plant Japanese Holly Fern on its own or in combination with shrubs or other perennials. Combine it with Cast Iron Plant or Boxwoods. Use it as a foundation plant where you don't want plants to get too tall.

# AUTUMN FERN
*Dryopteris erythrosora*

Autumn Fern is evergreen, adaptable, and easy to grow. It gets its name from the new growth that starts out a bronzy red before it turns green. Individual plants become quite large and wide, almost like a small shrub or groundcover. The selection 'Brilliance' is noted for keeping its luster and the orange new growth.

**Size:** Autumn Fern grows 1½ to 2½ feet tall by 2 to 3 feet wide.

**Conditions:** Plant Autumn Fern in part shade or shade. Once established, it will tolerate some drought. Cut back any tattered fronds in early spring before new growth emerges. Divide clumps in early spring.

**Zones:** 5 to 8

**Uses and Companions:** Autumn Fern makes a good foundation plant in combination with Boxwoods or Hellebores. Plant it in mass for a groundcover in the woodland or to establish a border in the shade.

# OSTRICH FERN
*Matteuccia struthiopteris*

This large elegant fern makes a statement in the garden. Ostrich Fern has a vase-shaped habit with fronds that are narrow at the base but get wider as they get taller, resembling ostrich plumes. A native deciduous fern, it spreads by lateral underground runners. The tightly wound young ornamental fiddleheads (new growth) can be cooked and are considered a delicacy by some.

**Size:** Ostrich Fern matures to 3 to 6 feet tall by 2 to 3 feet wide.

**Conditions:** Plant this fern in part shade or shade. It will tolerate a sunny location if the soil is kept moist. Soil rich in organic matter is ideal. In the wild it favors riverbanks and sandy soils. A layer of leaf mulch (1 to 2 inches) gives it a boost.

**Zones:** 3 to 8

**Uses and Companions:** Plant Ostrich Fern next to a pond with Cardinal Flower, *Lobelia cardinalis*. Combine it with *Iris ensata*, Japanese Iris, or Hostas for a contrast in foliage.

# CINNAMON FERN
*Osmunda cinnamomea*

Cinnamon Fern is appealing at every stage. In early spring when the fiddleheads unfurl, it provides architectural interest. Then in summer, the green fronds add graceful texture. There are separate fertile and sterile fronds. The fertile fronds, which bear the spores for reproduction, are cinnamon colored (resembling a cinnamon stick) and the sterile fronds are green. The fertile fronds are most visible early in the season.

**Size:** Cinnamon Fern grows 4 to 5 feet tall by 2 to 3 feet wide.

**Conditions:** A moist soil that is rich in organic matter is ideal in part shade or sun. Transplant crowns in early spring or summer. They should be replanted at the existing soil level.

**Zones:** 3 to 10

**Uses and Companions:** Use Cinnamon Fern as a backdrop for other perennials like Hostas. Combine it with shrubs like Virginia Sweetspire, *Itea virginica* 'Henry's Garnet', or *Clethra alnifolia*, Sweet Pepperbush.

# ROYAL FERN
*Osmunda regalis*

Adaptable and elegant, Royal Fern makes a statement in the woodland with its erect fronds that can be as tall as 6 to 8 feet. Although the fronds are made up of leaves that are twice cut, the overall effect of this fern is bold. A native to the eastern United States, it is found growing in swamps and other

wet sites but will also grow in an average garden soil. A deciduous fern, in autumn the fronds may take on shades of yellow before they die back for winter.

**Size:** The average size is 2 to 5 feet but this fern can grow up to 10 feet tall.

**Conditions:** Moist soil and part sun are ideal but this fern will grow in wet or average soils in part sun to shade.

**Zones:** 2 to 10

**Uses and Companions:** Combine Royal Fern with Cinnamon Fern and Ostrich Fern for a dramatic display. It also works well with Hostas and Hellebores. For a colorful accent, plant it with *Lobelia cardinalis*, Cardinal Flower.

# CHRISTMAS FERN
*Polystichum acrostichoides*

When I walk through the woods in winter, patches of green Christmas Fern stand out against the mostly brown forest floor. This evergreen native is handsome, tough, and easy to grow. It also tolerates periods of drought.

**Size:** Christmas Fern grows 24 to 36 inches tall. It forms a fountainlike clump that increases in size over time.

**Conditions:** Christmas Fern will grow in shade or part shade in moist or dry soils. In sunny locations it needs a moist soil. Once established, it will tolerate periods of drought.

**Zones:** 4 to 9

**Uses and Companions:** Use Christmas Fern to control erosion or add evergreen foliage to a deciduous woodland. Plant it under shrubs to create a carpet that holds up in every season. Plant it with Daffodils so the fern foliage can mask the bulb foliage as it ripens.

# SOUTHERN SHIELD FERN OR SOUTHERN WOOD FERN
*Thelpteris kunthii*

A native, Southern Shield Fern is big and bold and easy to grow. The tall, 1-foot-wide, arching green fronds add graceful texture to the garden. In autumn the sturdy green fronds take on shades of bronze. Southern Shield Fern is also known as the Southern Wood Fern.

**Size:** Southern Shield Fern grows 3 to 4 feet tall and 3 to 4 feet wide. This fern spreads by underground rhizomes.

**Conditions:** Plant this fern in part shade or shade in a soil that is rich in organic matter. It will also tolerate sunny locations if the soil is kept moist. Once established, it will tolerate some drought.

**Zones:** 6 to 9

**Uses and Companions:** Southern Shield Fern makes a good choice to border a path in the woodland. Combine it with Hellebores, Hostas, and other shade-loving perennials.

NORTHERN
MAIDENHAIR FERN

# MORE FERNS
## FOR SHADE

*Adiantum pedatum* — Northern Maidenhair Fern

*Cyrtomium fortunei* — Japanese Holly Fern

*Dennstaedtia punctilobula* — Hay-Scented Fern

*Dryopteris goldiana* — Goldie's Wood Fern

*Dryopteris ludoviciana* — Southern Wood Fern

*Dryopteris marginalis* — Marginal Wood Fern

*Lygodium japonicum* — Japanese Climbing Fern

*Onoclea sensibilis* — Sensitive Fern

*Osmunda claytonia* — Interrupted Fern

*Polystichum polyblepharum* — Tassel Fern

# THE CONTAINER BOG GARDEN

Pitcher Plants, *Sarracenia*, which are native and carnivorous, have long fascinated gardeners and children alike. Not only are these bog plants beautiful, they are heat and cold tolerant, growing happily outside in a container or in the garden year-round. The pitchers appear after the unusual hooded flowers and, depending on the species, may last well into autumn. Some people dry the pitchers and use them in flower arrangements. Their colorful leaves attract flies, bees, and moths. Nectar located along the opening of the pitcher lures insects within; once they are inside, they are trapped and digested by the plant. Companions for Pitcher Plants include *Drosera capillaries*, Pink Sundew; *Dionaea muscipula*, Venus Flytrap; and *Zephyranthes atamasco*, Atamasco Lily.

If you want to create a bog container, here are a few tips:

- The container should have some drainage, so that water can seep out as it would naturally in the wild.
- The soil should be a 50:50 mix of coarse sand and peat moss.
- Adding a layer of charcoal can help drainage.
- The container should be 12 to 18 inches deep and 12 inches wide or wider.
- Most bog plants (especially Pitcher Plants) need full sun.
- Keep the soil wet but don't allow standing water.
- Pitcher Plants do not like chlorinated water. You can use collected rainwater as a way to water your plants.
- There is no need to fertilize.

BOG CONTAINER GARDEN WITH PITCHER PLANTS

DRYOPTERIS MARGINALIS
MARGINAL WOOD FERN

Groundcovers that thrive in full sun offer a viable alternative to turf, especially in areas that are difficult to access with a mower, such as steep slopes and between stepping stones. They also provide transitions from shrubs and trees to perennials and bulbs. Some, like Creeping Raspberry, are tough and durable, making them well suited for areas like the planting strip between the sidewalk and the street, often referred to as the "hell strip" because of the abuse plants are often subjected to in this location. Like Creeping Raspberry, there are a number of Sedums including *Sedum tectractinum*, the Bronze Sedum, and *Sedum rupestre* 'Angelina' that are evergreen and durable as groundcovers, but they will not hold up to lots of foot traffic.

Flowering groundcovers like *Veronica* 'Georgia Blue' and *Erigeron karvinskianus*, Mexican Daisy, make great edgers for the front of the border or between blocks of colorful perennials. Herbs, like some of the Thymes, many of which are also fragrant, are also effective as groundcovers. They work well between paving stones along a path or in front of taller border perennials.

From a creative standpoint, you can plant blocks of different groundcovers in large areas, creating a beautiful patchwork effect.

# HARDY ICE PLANT
*Delospema cooperi*

For full sun and dry soils, this succulent-looking perennial is an ideal groundcover although it does not like wet soils in winter. The shiny foliage, fingerlike and evergreen, looks good all year and its bright purple to deep rose flowers cover the plant throughout much of the summer. It makes a good candidate for container gardening, if you make sure to let it dry out between waterings. A low grower that spreads quickly, this carefree beauty offers a lot and demands little in return.

**Size:** Plants grow 2 to 4 inches high and flowers are 2 to 3 inches across. Individual plants easily spread to 2 feet.

**Conditions:** Good drainage is essential and full sun is ideal, but some afternoon shade is okay. Space plants 12 to 15 inches apart. Use a gravel mulch, and watch plants spread over the top.

**Zones:** 6 to 9

**Uses and Companions:** Let it cascade over a wall, or plant it in the front of other perennials in the flower border or rock garden. Combine it with other plants with low water requirements. For contrast, combine it with conifers, such as Blue Atlas Cedar.

# MEXICAN DAISY OR FLEABANE
*Erigeron karvinskianus* 'Profusion'

This woody perennial blooms all summer with tiny white daisies that fade to pink as they age. It's also called Fleabane due to the look of the flowers. This is a great filler in sunny borders. The green foliage looks delicate, but plants are tough and tolerate periods of drought once established. Mexican Daisy is a good choice for its long season of bloom and lack of fussiness.

**Size:** Plants grow 8 to 10 inches tall and up to 2 feet wide. Individual flowers are ½ to ¾ inches wide.

**Conditions:** This plant prefers full sun to part shade and moist, well-drained soil. Wait until early spring to cut plants back, just before new growth begins.

**Zones:** 6 to 9

**Uses and Companions:** Use Mexican Daisy in mass for a groundcover between other larger perennials like Coneflowers and Iris or shrubs in the border. Combine it with foliage perennials like Lamb's Ear. It also makes a good subject for containers with larger-flowered plants or between stepping stones.

# CREEPING ST. JOHNSWORT

*Hypericum calycinum*

An aggressive grower, this *Hypericum* spreads by underground roots, quickly covering large areas and forming a dense groundcover. These traits make *Hypericum* a good choice to control erosion on slopes, or as a groundcover in areas where grass won't grow because of poor soil. The foliage color is yellow-green in the shade and green in the sun. Once established, it tolerates periods of extreme heat and drought.

**Size:** Growing to only 1 foot high, individual plants easily spread to 2 feet or wider. The leaves are 4 inches long.

**Conditions:** For the best results, plant in part sun in average garden soil, although it grows in a wide range of soil types. In full sun plants need plenty of moisture. The foliage is mostly evergreen, but if it becomes tattered, prune back hard in early spring.

**Zones:** 4 to 8

**Uses and Companions:** It makes a handsome edger at the front of a border. Combine it with Coneflowers, Salvia, and Butterfly Bush.

# CREEPING PHLOX OR THRIFT

*Phlox subulata*

Driving along country roads in early spring, it's hard to miss Creeping Phlox when it blooms. Brilliant flowers of pink to rose, lavender, or white cover roadside banks and drainage ditches. Creeping, evergreen, needlelike foliage forms thick, ground-hugging mats. In or out of bloom, it provides a carefree evergreen carpet. Another common name for this Phlox is Thrift.

**Size:** Plants grow 6 inches high with a spread of 2 feet per plant.

**Conditions:** This mat-forming Phlox thrives in full sun and average garden soil. Make sure the soil is well drained and not too rich. Propagate clumps by division after plants finish flowering in spring.

**Zones:** 3 to 9

**Uses and Companions:** Plant this phlox on banks, at the front of the flower border, between stepping stones, and in the rock garden with other perennials.

## GRO-LOW SUMAC
*Rhus aromatica 'Gro-low'*

This native deciduous shrub is ideal for covering poor or dry soils and on slopes or banks. Fast growing and adaptable, the species is also known as Fragrant Sumac for its aromatic foliage when crushed. In spring it is covered with small yellow flowers. The lobed leaves are shiny and green all summer. In autumn they light up the landscape when they turn shades of red and orange. Once established, this Sumac will tolerate periods of drought.

**Size:** This plant is truly shrubby, growing only 2 to 3 feet tall but up to 9 feet wide.

**Conditions:** Full sun or part shade and average, well-drained soil are best, but like other Sumacs, it will also tolerate poor soils. The branches often root where they touch the soil, so this plant will quickly form a thicket.

**Zones:** 4 to 9

**Uses and Companions:** Gro-low Sumac is great for steep, sunny banks where it is hard to establish plants. Combine it with ornamental grasses and Asters for a colorful fall display. This plant is ideal for informal and native gardens.

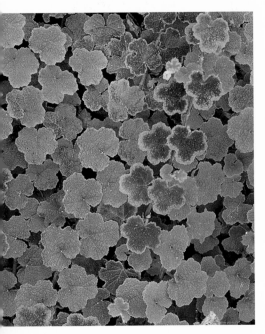

## CREEPING RASPBERRY
*Rubus pentalobus* or *R. calycinoides*

Tough and durable, this evergreen ground-cover grows happily in a range of conditions. The wrinkled, thick green leaves have wavy edges and look good throughout the growing season. With the arrival of cooler weather in autumn, they take on shades of red and orange. This is a perfect plant for the strip between the sidewalk and the street where other plants often languish. Once established, Creeping Raspberry is fairly drought tolerant. It's a good choice to help suppress weeds too.

**Size:** Creeping Raspberry grows 6 to 10 inches tall and easily 5 feet wide in three years.

**Conditions:** The ideal conditions are full sun or part shade in average garden soil, provided it is well drained. Creeping Raspberry is easy to propagate; just dig up sections where the stems touch the soil and root.

**Zones:** 6 to 9

**Uses and Companions:** Creeping Raspberry makes a handsome evergreen groundcover in combination with Daffodil bulbs. Plant it on a sunny slope or at the edge of a shrub border.

# CREEPING SEDUM

*Sedum rupestre* 'Angelina'

This evergreen Sedum has chartreuse foliage that brightens the garden year-round. It also provides interesting texture with its needlelike leaves. Creeping Sedum stays low to the ground but spreads out quickly. It makes for great contrast when it's planted with broad-leaved plants like *Canna* and *Alocasia*.

**Size:** Creeping Sedum grows 4 inches tall with a spread to 2 feet in a single growing season.

**Conditions:** Full sun and well-drained soil are ideal, but this Sedum will also tolerate poor soils.

**Zones:** 3 to 8

**Uses and Companions:** Plant it in the perennial border with Coneflowers, Black-eyed Susans, and Lamb's Ears. It also makes a good choice for the rock garden with hardy Cacti and succulents. Use it as an edger in the mixed border.

# BRONZE SEDUM

*Sedum tectractinum*

This Sedum hails from China but is very happy in Southern gardens, where it is evergreen. The round, succulent-like leaves vary in size from a penny to a nickel. It gets its common name from the bronze tinges on the tips of the leaves. In summer it is covered with bright yellow flowers, adding to its appeal. It forms a thick groundcover in no time, but will not take over the garden. You can plant it between paving stones in a patio. Once established, it will tolerate periods of drought.

**Size:** Bronze Sedum grows only about 4 inches tall but easily spreads to 2 feet wide.

**Conditions:** Ideal conditions include full sun or part shade and well-drained soil. This Sedum forms somewhat of a mound as it ages. Propagate plants by division in spring and fall.

**Zones:** 4 to 9

**Uses and Companions:** Use it as a groundcover in the perennial border. Combine it with Euphorbias, Coneflowers, and other Sedums. In the rock garden, let it cascade over walls, or use it to cover slopes. Combine it with annuals in containers for summer color.

# THYME: GOLDEN, GOLDEN LEMON, AND WOOLY

*Thymus vulgaris* 'Aureus', also *Thymus pulegioides* 'Goldentime' or *Thymus* x *citriodorus* 'Aureus', *Thymus lanuginosus*

In addition to being a popular culinary herb, Thyme makes a great groundcover. Some, like Woolly Thyme, *T. lanuginosus*, hug the ground and tolerate moderate foot traffic. While the names are often debated and confused, the plants are good performers for hot, dry sites in sandy soils. For fragrant, year-round beauty they are hard to beat. Deer do not favor Thymes!

**Size:** Thyme grows 2 to 8 inches tall depending on the particular species. Most will spread out to 2 feet wide or more. While they vary in height and habit, all are fragrant, and most are evergreen.

**Conditions:** Full sun and well-drained soil are ideal. If patches die out because of humidity, just pull away the dead part, and new growth will fill in.

**Zones:** 5 to 9

**Uses and Companions:** Thyme is great for growing in containers with other herbs like Rosemary or Lavender, or as a groundcover between stepping stones. It also makes a good companion for *Viola* or Pansies in winter.

# CREEPING SPEEDWELL

*Veronica peduncularis* 'Georgia Blue'

This carefree, creeping *Veronica* species is covered with tiny, bright blue flowers with a white eye that last for weeks in early spring. If you're lucky it will also bloom intermittently during the summer. The glossy foliage starts out with red tinges in spring and then turns green for summer. In autumn it takes on bronze tones. *Veronica* 'Georgia Blue' attracts hummingbirds and butterflies to the garden.

**Size:** Creeping Speedwell grows 8 to 12 inches high and wide.

**Conditions:** Full sun to part shade and well-drained soil are ideal, but it will tolerate a range of soil types. To keep plants tidy, shear them back by half after they finish blooming.

**Zones:** 5 to 9

**Uses and Companions:** This groundcover is perfect for the front of the border as an edger or under shrubs like Roses and Hydrangeas. Combine it with other perennials like Lamb's Ear, Iris, and Hardy Geraniums. Let it cascade over walls in the rock garden, where it can act as a groundcover for bulbs like summer-blooming Alliums.

LAMB'S EAR

# MORE
# GROUNDCOVERS
## FOR SUN

*Acorus gramineus* 'Minimus Aureus' *— Dwarf Golden Sweet Flag*

*Carex buchananii* *— Leatherleaf Sedge*

*Carex flacca* 'Blue Zinger' *— Blue Zinger Sedge*

*Dianthus* 'Bath's Pink' *— Cheddar Pinks*

*Nepeta* 'Walker's Low' *— Catmint*

*Sedum kamtschaticum* *— Orange Sedum*

*Sedum makinoi* 'Ogon' *— Creeping Sedum*

*Sedum spurium* *— Two-Row Stonecrop*

*Stachys* 'Helen Von Stein' *— Lamb's Ear*

*Veronica repens* 'Sunshine' *— Sunshine Dwarf Veronica*

# ORNAMENTAL AND EDIBLES

Combining ornamental and edible plants or growing edible plants that are also ornamental is not a new concept, but how it is implemented varies greatly.

For some gardeners, growing plants is about "putting food on the table." But for many of us it is exciting to grow plants that are not only edible but add their own unique beauty to the landscape. While space may limit some from including a great variety of edibles, there are a number of trees, shrubs, and herbaceous plants that offer both beauty and tasty fruits too. Many of these are easy to grow and will thrive in our Southern gardens.

Here is list of ornamentals with edible fruits:

JUJUBE

*Actinidia arguta* 'Issai'—Hardy Kiwi Vine
  (you'll need a male and female plant
  for fruit production)
*Actinidia deliciosa* 'Hayward'—Kiwi
*Asimina triloba*—Paw Paw
*Cudrania tricuspidata*—Che Fruit
*Diospyros kaki* 'Fuyu'—Asian Persimmon
  (many selections)
*Feijoa sellowiana*—Pineapple Guava
*Ficus carica* 'Brown Turkey', 'Celeste', or
  'Black Mission'—Hardy Fig
*Malus domestica* cultivars—Apple
  (many selections thrive in the South)
*Punica granatum*—Pomegranate
*Vaccinium ashei* 'Premiere', 'Climax', or
  'Tifblue'—Rabbiteye Blueberry
  (plant two different varieties to ensure
  the best fruiting )
*Ziziphus jujube*—Jujube

For more information contact the North American Fruit Tree Explorers at www.nafex.org.

# GROUNDCOVERS
FOR SHADE

Often as gardens mature, they change. What was once a garden in full sun is now in shade, creating a need for covering bare spots with plants. Groundcovers that thrive in the shade are useful under shrubs, trees, and perennials, or between stepping stones, often in places where growing turf is impractical. Groundcovers are also useful for creating an evergreen carpet under spring bulbs.

Some, like Green and Gold, are well suited for the woodland garden and will quickly cover a steep slope. Others, like Dwarf Mondo, *Ophiopogon japonicus* 'Nana', which only grows 3 inches tall, offer an alternative to turf and will tolerate moderate foot traffic. Variegated Blue Lilyturf, *Liriope muscari*, is ideal for lining paths, edging beds, or for replacing large areas where lawn once grew.

Golden Creeping Jenny has handsome chartreuse foliage and will tolerate damp soils. Also tough and colorful are the numerous selections of Common Bugleweed.

In this chapter, I have not included English Ivy because it is invasive and difficult to get rid of once you have it in your garden. A small patch can spread quickly growing out and up and smothering other plants. Instead, I offer alternatives that I believe are effective and less problematic for our Southern gardens.

# COMMON BUGLEWEED

*Ajuga reptans*

Bugleweed is a vigorous evergreen groundcover that spreads quickly by runners to form a mat of foliage. Most types hug the ground and make an ideal carpet under shrubs or in combination with bulbs. The 4- to 5-inch spikes of blue to purple flowers (there is also a form with white flowers) are a bonus and appear from spring until early summer. Depending on the selection, the foliage can be green, bronze, a blend of colors, or variegated. Although Bugleweed has many uses, it is not a good choice for high-traffic areas.

**Size:** 'Chocolate Chip' has narrow leaves that are a mix of chocolate, purple, and green; it grows 4 inches high with a spread of 8 to 10 inches. 'Catlin's Giant' has 6-inch-long purple-and-green leaves and flower spikes that are 8 inches tall. It is more of a clumper.

**Conditions:** Half shade is idea for Bugleweed, especially selections with colorful foliage, but they will tolerate full shade or full sun if they receive enough moisture.

**Zones:** 3 to 10

**Uses and Companions:** Bugleweed is great for spots under trees, shrubs, and with bulbs. It is also good for planting on banks to control erosion or between stepping stones.

# GREEN AND GOLD

*Chrysogonum virginianum*

With its bright yellow flowers and green foliage, Green and Gold stands out on the woodland floor in mid- to late April. It grows 6 to 9 inches tall and spreads easily and quickly. The ray flowers are somewhat reminiscent of stars and appear in spring and fall with the odd flower in summer too. 'Eco Laquered Spider' can spread 3 feet or more in two years by runners, which put out roots wherever they touch soil.

**Size:** The leaves are 1 to 3 inches long and the flowers are 1 to 1½ inches long. This native spreads quickly, especially 'Eco Laquered Spider'.

**Conditions:** Green and Gold will grow in shade or part shade. 'Eco Laquered Spider' will also tolerate full sun, provided it receives enough moisture. While well-drained soil is ideal, this perennial will not tolerate long periods of drought.

**Zones:** 4 to 9

**Uses and Companions:** This groundcover can be aggressive but it is also easy to pull out wherever you don't want it. Plant it in the rock garden, woodland, or under shrubs in the shade.

# SPOTTED NETTLE

*Lamium maculatum*

If you need an aggressive groundcover, consider Spotted Nettle. This trailing perennial grows 6 inches tall and spreads 2 to 3 feet. The selections are less weedy than the species and offer handsome foliage like 'Beacon Silver', with spikes of pink flowers and green-edged, silver-gray leaves, or 'White Nancy', also with green-and-gray variegated foliage, but the flowers are white.

**Size:** Spotted Nettle grows 6 to 12 inches tall with a 2- to 3-foot spread.

**Conditions:** This adaptable plant will grow in full sun or part shade in an average garden soil. It will also tolerate some drought. It is mostly evergreen except in the coldest winters.

**Zones:** 3 to 9

**Uses and Companions:** Plant Spotted Nettle so it trails over a wall or grows under shrubs. Plant it under trees and shrubs where grass won't grow. Combine it with dark-leaved Hostas and ferns.

*Liriope* is also known as Monkey Grass or Lilyturf. This tough evergreen perennial forms large clumps of plants that look like grass with leaves that are ½ inch wide and up to 2 feet long. While the clumps increase in size quickly, this species of *Liriope* does not spread by underground runners, so it is easy to control. An undemanding plant, it makes an effective groundcover or edger. One thing to note: it does not like wet feet. The spikes of lavender or white flowers in summer add to its charm.

# VARIEGATED BLUE LILYTURF

*Liriope muscari* 'Variegata'

**Size:** Variegated Blue Lilyturf grows 12 to 15 inches tall and forms clumps up to 2 feet across. Flowers appear on spikes from 4 to 12 inches high.

**Conditions:** Variegated Blue Lilyturf grows in average garden soil. Full shade or filtered sun is best. It will tolerate periods of drought. Cut back tattered foliage in late winter before new growth starts.

**Zones:** 6 to 10

**Uses and Companions:** Variegated Blue Lilyturf makes an effective groundcover, edger for borders or flowerbeds, or a mass planting in the shade. It also works well in areas where plants must tolerate traffic from humans and animals. Plant Variegated Blue Lilyturf under trees and shrubs where root competition can be a problem.

# GOLDEN CREEPING JENNY

*Lysimachia nummularia* 'Aurea'

Not only is this plant easy to grow, forming roots wherever the leaf joints touch the soil, it offers handsome small, rounded, chartreuse foliage for shade or part sun. And it grows in soils that are wet or dry and tolerates moderate foot traffic. In the summer it produces yellow flowers about 1 inch across.

**Size:** Golden Creeping Jenny forms long runners to about 2 feet long that quickly fill up large areas. This ground hugger is only 2 to 4 inches tall.

**Conditions:** Plant Golden Creeping Jenny in an average garden soil in part or full shade. It will tolerate both wet and dry soils.

**Zones:** 4 to 8

**Uses and Companions:** Site this perennial along the edge of ponds, between stepping stones, or in combination with Hostas and ferns. It also looks good when planted with Cardinal Flower, *Lobelia cardinalis,* and its spikes of intense red flowers in August.

# MAZUS

*Mazus reptans*

*Mazus* looks delicate, but when planted between stepping stones, it will tolerate a moderate amount of foot traffic. This ground hugger grows 1 to 2 inches tall and spreads by way of slender stems with semi-evergreen leaves that are almost oval, putting down roots wherever they touch soil. The tiny snapdragon-like flowers that appear in late spring to early summer, single or in clusters of two to five, are purple-blue with white and yellow markings. There is also a white flowered form. It is evergreen except in the coldest winters.

**Size:** Growing 1 to 2 inches tall, *Mazus* spreads out as far as you let it.

**Conditions:** *Mazus* will grow happily in full sun, part shade, or shade. Make sure to keep the soil moist.

**Zones:** 5 to 9

**Uses and Companions:** *Mazus* is a great groundcover under shrubs over bulbs or in combination with ferns, Hostas, and other woodland treasures. Plant it at the edge of a pond, near rocks, or between paving stones where foot traffic won't bother it.

# MONDO GRASS

*Ophiopogon japonicus*

Mondo Grass forms evergreen clumps of grass that spread by underground stems. Although it is sometimes slow to establish, with its dark green leaves that are only ⅛ inch wide, it provides a choice alternative for a traditional lawn in shady spots where grass won't grow.

**Size:** Mondo Grass is 6 to 8 inches tall but there are a number of dwarf selections that only reach 3 inches high at maturity. Dwarf cultivars include 'Kyoto Dwarf' and 'Nana'.

**Conditions:** Filtered sun and moist, well-drained soil are ideal, but Mondo Grass will tolerate full shade. If foliage gets ragged, cut it back in early spring before new growth starts.

**Zones:** 6 to 10

**Uses and Companions:** Mondo Grass works well between steppingstones or as an edger for a path in the woodland. It will tolerate moderate foot traffic. Plant it under shrubs or combine it with ferns and Hostas in the shade.

# WHORLED STONECROP

*Sedum ternatum*

This native evergreen perennial occurs in open woodlands in the eastern United States. This is truly a plant that provides four seasons of interest. It grows in a range of sites including moist soil with lots of organic matter and dry rocky slopes that are also subject to wind. Because of its succulent leaves, it will tolerate periods of drought, making it a choice plant for areas that dry out periodically. The ½-inch-wide, starlike, white flowers are a bonus in spring, occurring in flat-topped clusters of two to four flowers.

**Size:** This *Sedum* grows up to 10 inches tall with leaves that are ¼ to ¾ inch long and across. Mature plants spread out 4 to 8 inches or more.

**Conditions:** Plants will grow in sun or shade. If you choose a sunny spot, make sure plants receive lots of moisture.

**Zones:** 4 to 8

**Uses and Companions:** Whorled Stonecrop makes an effective groundcover on its own, on rocky slopes, or the woodland floor. Combine it with ferns and wildflowers like *Iris cristata* or the hardy *Geranium maculatum*. This groundcover is reported to be deer resistant, good news for gardeners who face this challenge.

# LARGE PERIWINKLE
*Vinca major*

If you need quick cover in the shade, especially on a slope or embankment, this aggressive trailing groundcover is a good choice. An evergreen, the leaves of Large Periwinkle are 2 to 3 inches long and the plant grows 1 to 2 feet high. Single purple flowers appear in the spring, adding to its charm. There is also a variegated selection that works well in the ground or in decorative containers. A word of caution: all *Vinca* varieties are poisonous if ingested.

**Size:** Large Periwinkle grows 1 to 2 feet high, spreading out indefinitely.

**Conditions:** Average, well-drained soil in full shade or part sun is ideal. If you plant Large Periwinkle in full sun, make sure it is watered regularly. If plants lack vigor, shear them to several inches high in early spring to encourage growth. This perennial is easy to propagate by cuttings or division.

**Zones:** 7 to 9

**Uses and Companions:** Use Large Periwinkle to control erosion on a bank or slope. You can also plant it in combination with Daffodils or Summer Snowflakes; the foliage will hide the bulb foliage after bulbs bloom in spring.

Common Periwinkle is more refined than its close relative, Large Periwinkle. The smaller, glossy leaves and more restrained habit make it suitable for both formal and informal gardens. There are a number of different selections with flowers that range from white to various shades of blue. *Vinca minor* 'Bowles Variety' offers leaves that are somewhat larger and deep blue flowers. A rapid spreader, it puts out roots at leaf nodes when they touch the soil.

# COMMON PERIWINKLE
*Vinca minor*

**Size:** This smaller Periwinkle has leaves that are less than 2 inches long, flowers that are 1 inch across, and plants that reach 6 inches high at maturity.

**Conditions:** Plant this Periwinkle in full shade or part sun in an average garden soil. Cut back plants in early spring if plants need rejuvenating.

**Zones:** 4 to 9

**Uses and Companions:** Plant evergreen Periwinkle as a foundation plant with ferns and small shrubs. Combine it with early Daffodils for spring bloom.

*CAREX 'EVERGOLD'*

# MORE GROUNDCOVERS
## FOR SHADE

*Acorus gramineus* 'Ogon'—*Dwarf Sweet Flag*

*Aegopodium podagraria* 'Variegata'—*Variegated Goutweed*

*Carex morrowii* var. *temnolepis* 'Silk Tassel'—*Sedge*

*Carex oshimensis* 'Evergold'—*Sedge*

*Carex plantaginea*—*Plantain-leafed Sedge*

*Carex siderosticha* 'Variegata'—*Creeping Broad-leafed Sedge*

*Galax urceolata*—*Wandflower*

*Pulmonaria angustifolia*—*Blue Lungwort*

*Selaginella uncinata*—*Peacock Moss*

*Senecio aureus*—*Golden Ragwort*

# ORNAMENTAL GRASSES

BIG BLUESTEM

Ornamental grasses and related grasslike plants including sedges and rushes offer long seasonal interest in the garden and require minimal care.

They provide color, texture, form, and add grace, beauty and movement to the landscape. There are grasses for sun or shade, as well as those that are drought friendly, and those that tolerate damp soils. Before you plant, do a little research, and be aware that some grasses may be aggressive or may reseed in the garden, creating potential problems down the road. Remember, it is always a matter of "the right plant for the right place."

Consider the following varieties for your garden.

**Big Bluestem**
*Andropogon gerardii*
SUN
A clumper, it grows 5 to 8 feet tall, and its blue-green summer foliage turns rich orange or copper red in fall.

**Feather Reed Grass**
*Calamagrostis* × *acutiflora* 'Karl Foerster'
SUN or PART-SHADE
This upright clumper grows 6 feet tall with deep green foliage; in late summer, vertical buff-colored plumes emerge.

**Sedge**
*Carex morrowii* var. *temnolepis* 'Silk Tassel'
SUN WITH MOIST SOIL or SHADE
This Sedge has dark green leaves with a white margin that are only ⅛ inch wide. It grows 1 foot tall by 2 feet wide.

**Sedge**
*Carex oshimensis* 'Evergold'
SHADE
Clumps of variegated foliage, dark green margins with a wide creamy-white strip in the middle, mark this distinctive plant. It grows 16 inches tall by 2 feet across.

CAREX 'TOFFEE TWIST'
WITH COLEUS AND NEW
GUINEA IMPATIENS

Annual vines offer the opportunity to experiment with color, fragrance, and texture. Many bloom from late spring until frost, providing a long season of interest. Vines afford gardeners an easy way to jazz up the landscape without committing to a particular plant or style beyond one growing season. Annual vines such as Mexican Flame Vine add bold color and will happily mix and mingle with other plants, while Mandevilla needs only a mailbox or post to make a statement all its own. Still others, such as Cypress Vine, a twining climber with delicate fernlike foliage and scarlet blooms, will spread out like a groundcover it you don't give it something to grow up and on.

Some, like Purple Hyacinth Bean, produce beautiful seedpods that add interest to the fall and winter garden (the seeds inside the pods provide the source for next year's vines), while others like Moonflower fill the air with their sweet scent.

Annual vines trained on a structure provide an easy way to add height to an herbaceous border. Whether you grow annual vines on their own or in combination with perennial vines, they require minimal care and offer much in return.

## COMMON ALLAMANDA
*Allamanda cathartica*

This tropical vine brightens the landscape with large, golden trumpets, 5 inches wide by 3 inches long. Its glossy green foliage adds to its charm. Common Allamanda quickly reaches great heights but makes a large shrub if it is pinched back regularly. Bush Allamanda, *Allamanda schottii*, is more compact with smaller flowers that are yellow with an orange or reddish tint. It can also be trained as a vine or small shrub for containers. Both provide welcome color in the summer garden.

**Size:** A rapid grower, *A. cathartica* can easily reach 6 to 8 feet or taller with a support.

**Conditions:** Full sun and moist, well-drained soil provide the ideal environment. Tie up the vines if you want it to grow up. It can also be used for hanging baskets or allowed to cascade over a wall.

**Zones:** 9 to 10 (tropical)

**Uses and Companions:** This tropical puts on a show during hot summer months. Grow it as a specimen or combine it with other annuals like Coleus, Plumbago, and Pentas. Use it in containers, window boxes, and hanging baskets, or train to a trellis or arbor.

## PURPLE HYACINTH BEAN
*Dolichos lablab purpureus*

With a great botanical name and an interesting history, this colorful vine is very easy to grow and provides quick cover in no time. It has been reported that Thomas Jefferson grew this twining vine at Monticello. Still popular today, it produces purple or white sweet pea-like flowers with purple and green stems. The foliage is green and wine colored. The shiny pods that appear from late summer into fall are a dark magenta color. Both the blooms and pods can be used as cut flowers for arrangements.

**Size:** This tropical vine can easily grow to 15 to 20 feet or more. The pods are 2½ inches long.

**Conditions:** Once established, this vine is drought tolerant. Plant in full sun or part shade in well-drained soil.

**Zones:** 9 to 10

**Uses and Companions:** This vine is a good choice where quick cover is needed on a chainlink fence or other structure. Let it climb up an arbor or pergola. Leave the seedpods until the first frost for late-season interest in the garden.

# MOON VINE

*Ipomoea alba*

If you enjoy intoxicating fragrance, Moon Vine is a must for your garden. This fast-growing vine has heart-shaped leaves and large, white, sweetly scented flowers, up to 6 inches across. The blooms open in late afternoon and usually close up the following morning, although on overcast days they may stay open longer. Perfect for the evening garden, Moon Vine can be easily trained to cover a fence, arbor, or trellis.

**Size:** This vine grows 20 to 30 feet in one season. The leaves are 3 to 8 inches long.

**Conditions:** Plant Moon Vine in full sun in well-drained soil.

**Zones:** 9 to 10

**Uses and Companions:** You can train Moon Vine to grow on an arbor with a Climbing Rose so that when the Rose has finished blooming, Moon Vine will take over, filling the air with its perfume. Train it up a lattice to create a summer screen or let it ramble over the mailbox. Combine it with other vines like Cypress Vine, which has red flowers.

# CYPRESS VINE

*Ipomoea quamoclit*

This delicate-looking vine has scarlet flowers (there is also a white form) and fernlike foliage. Although it looks dainty, it is an aggressive grower that will act like a groundcover if it doesn't have a structure to grow on. The tubular flowers flare at the mouth like a five-pointed star. Great for attracting hummingbirds to your garden, this old-fashioned plant is also called Cardinal Climber.

**Size:** A twining vine to 20 feet or taller. The finely divided leaves, 2½ to 4 inches long, are reminiscent of a spider web. The small flowers are 1½ inches long.

**Conditions:** Plant this twining vine in full sun or part shade in well-drained soil.

**Zones:** 9 to 10

**Uses and Companions:** Let it sprawl over the edge of a wall or containers with other annual grasses and flowers such as Pentas, Coleus, and Plumbago. Let it twine its way over structures like a trellis, fence, or arbor. It can be used as a groundcover in sunny spots with other annuals.

# MANDEVILLA HYBRIDS

*Mandevilla* spp. and hybrids

For abundant summer blooms from spring until frost, this tropical vine from Brazil is hard to beat. It does well in containers where it can ramble and scramble, or it can be trained up a support like a trellis or arbor. Long popular is *Mandevilla* 'Alice du Pont' with rich pink bell-shaped flowers 2 to 4 inches wide and shiny, leathery foliage. The Sun Parasol™ series offers brightly colored flowers that don't fade in the sun and heat and includes 'Sun Parasol Dark Red', 'Sun Parasol Crimson', 'Sun Parasol Giant Crimson', 'Sun Parasol Pink', 'Sun Parasol Cream Pink', and 'Sun Parasol Giant Pink'.

**Size:** Mandevilla's height depends on the selection. *M.* 'Alice du Pont' grows 20 to 30 inches tall and produces flowers 2 to 4 inches wide. Sun Parasol™ varieties also vary, ranging 6 to 15 feet tall, with blooms 4 to 6 inches wide.

**Conditions:** Full sun and moist, well-drained soil are best. It benefits from a dose of water-soluble fertilizer every 2 to 3 weeks or slow-release fertilizer when you plant it.

**Zones:** 9 to 10

**Uses and Companions:** Use this vine in hanging baskets, containers, or in the ground. Train it up a trellis, arbor, or fence. Combine it with other annuals like Lantana, Coleus, Plumbago, or the chartreuse colored Sweet Potato Vine.

# SPANISH FLAG

*Ipomoea lobata*

Grown in gardens for more than 100 years, this vine is also known as Exotic Love or Firecracker Vine. It blooms from midsummer until frost with arching sprays of tubular flowers that open crimson and then turn orange, yellow, and white. The effect is very colorful with all three colors appearing at once. The segmented, three-part leaves are interesting in their own right. Hummingbirds have been sighted visiting these plants.

**Size:** Up to twelve flowers per stem occur on this vine, which twines 15 to 20 feet tall.

**Conditions:** Plant in full sun to part shade in well-drained soil.

**Zones:** 9 to 10

**Uses and Companions:** Whether you use this vine in a hanging basket or train it to grow on a trellis or other structure, it looks great on its own or with other annuals. Combine it with red and blue Salvias, Lantana, and Mexican Sunflowers.

# PASSION FLOWER

*Passiflora × alatocaerulea and P. coccinea*

Passion Flower is not only a beautiful vine but it's a source of food for caterpillars of the Gulf fritillary butterfly. A rampant grower, it will take over a field if left to its own devices and not trained up a structure. The unique flowers resemble a lacy crown and to some represent a halo or crown of thorns, hence the name Passion Flower.

**Size:** This vigorous vine produces long stems up to 25 feet long and has tri-lobed leaves that are 3 inches long. The sterile blue flowers are 2 inches across. Red Passion Flower, *Passiflora coccinea*, produces bright scarlet blooms that are 3 to 5 inches across and edible fruits that turn yellow or orange when ripe.

**Conditions:** Passion Flower tolerates a wide range of soil types. Full sun and moderate water are all these plants need to thrive.

**Zones:** 9 to 12

**Uses and Companions:** Train Passion Flowers to grow up a trellis or arbor or let them sprawl like groundcovers. Combine it with Lantana, ornamental grasses, Coleus, and Salvias.

# MEXICAN FLAME VINE

*Pseudogynoxys chenopodioides*

Bright orange-red daisylike blooms with golden centers against dark green leaves make this vine a standout in the summer garden from late spring through fall. With its well-behaved manner it makes a good candidate to combine with hardy succulents like Agave and Sedums or other perennials and annuals. Use it in the hot border with Purple Smoke Tree and Red Hot Pokers.

**Size:** This twining plant grows 8 to 10 feet tall with 2-inch-wide blooms and coarsely toothed, light-green leaves, 1 to 4 inches long and ½ to 1 inch wide.

**Conditions:** Full sun or light shade is best with well-drained soil.

**Zones:** 9 to 10

**Uses and Companions:** Combine these bright blooms with gray foliage plants such as Lamb's Ears and Agaves. Train it to climb up a trellis support or use another plant for a living arbor. Plant it in a container with other annuals or let it trail over a wall.

# VARIEGATED POTATO VINE

*Solanum jasminoides* 'Variegata'

This is a vine to grow for its handsome bright green and gold foliage; the small white flowers are a bonus. Well behaved and easy to control, it has also been reported as being deer resistant. A twiner, Variegated Potato Vine is attractive for months in the garden and requires very little attention. Trained up an arbor, it provides a canopy of light shade and interesting texture.

**Size:** Variegated Potato Vine can grow 15 to 20 feet long.

**Conditions:** Plant in full sun or part shade and well-drained soil.

**Zones:** 9 to 11

**Uses and Companions:** Combine this with large flowered Clematis for a stunning effect. Variegated Potato Vine also looks good with Lantana, red and blue Salvias, and purple Verbena.

# BLACK-EYED SUSAN VINE

*Thunbergia alata*

For nonstop colorful flowers spring through fall, this vigorous heat-and-humidity tolerant vine is a top contender. The distinctive triangular leaves are fuzzy and wrinkled. This vine attaches to structures and other plants with tendrils that grab. The selection 'Sunny Lemon Star' offers bright-yellow flowers with a black eye and 'Sunny Orange Wonder' offers brilliant orange flowers. A white-flowered selection is also worth mention.

**Size:** This vine produces stems that are 8 to 10 feet long and plants that are 2 to 4 feet across.

**Conditions:** Full sun or light shade and well-drained soil are ideal.

**Zones:** 9 to 10

**Uses and Companions:** Plant this vine in hanging baskets or train it to grow on a trellis, arbor, or other structure. In a pot it looks good with other bright flowers such as *Salvia* or Lantana.

MORNING GLORY

# MORE ANNUAL
## VINES

*Antigonon leptopus* — Coral Vine

*Asarina scandens* — Creeping Gloxinia

*Cardiospermum halicacabum* — Love-in-a-Puff

*Clerodendrum thomsoniae* — Bleeding Heart Vine

*Ipomoea purpurea* — Morning Glory Vine

*Ipomoea tricolor* 'Heavenly Blue' — Heavenly Blue Morning Glory

*Lathyrus odoratus* — Sweet Pea

*Pandorea jasminoides* — Bower Vine

*Phaseolus coccineus* — Scarlet Runner Bean

*Rhodochiton atrosanguineum* — Purple Bell Vine

PROVEN PLANTS

95

# FABULOUS FRAGRANCE

How lovely it is to inhale the perfume of Daffodils in spring or Roses in summer. Whatever the season, each offers its own intoxicating scents. Sometimes when we enter a garden, we notice the fragrance of a flower long before we discover where it is planted. Even the tiniest blossoms can fill the air with sweetness like those of Sweet Box, *Sarcococca hookerana humilis*, whose flowers are hidden by its foliage.

Fragrance in plants exists to attract pollinators to the blooms, but for humans scents can trigger strong reactions and emotions. Gardening with fragrant plants adds another dimension to your landscape and encourages visitors to "stop and smell the flowers." While there are many fragrant flowers, fragrance can be found in foliage, bark, and seeds too. In some cases, as with many herbs, the leaves must be crushed or bruised to release their scent.

When designing a garden for fragrance, site plants where their aromas will be appreciated, such as along pathways or near entryways.

**Favorites for Fragrance**

*Buddleia davidii* — Butterfly Bush
*Calycanthus floridus* — Sweetshrub
*Daphne odora* — Winter Daphne
*Gardenia jasminoides* — Gardenia
*Hedychium coronarium* — Butterfly Ginger
*Lilium formosanum* — Formosa Lily
*Narcissus* species and cultivars — Daffodils
*Osmanthus fragrans* — Tea Olive
*Rosa* species and cultivars — Roses
*Rosmarinus officinalis* — Rosemary

*PHASEOLUS COCCINEUS*
SCARLET RUNNER BEAN

# PERENNIAL
## VINES

Vines offer an opportunity to take advantage of the vertical spaces in your garden. Whether they ramble, scramble, cling, or twine, there is always room for versatile vines. When selecting a spot, thought should be given to the habit of the vine and how big it will be when it matures, as well as what it will look like in winter. The method by which individual vines attach themselves, either to another plant or a surface, should be taken into consideration. Deciduous vines like Climbing Hydrangea use aerial roots to attach themselves, while vines like *Wisteria frutescens* 'Amethyst Falls' will initially need to be tied, but once they get started, they will continue to grow up the structure or plant you are training them on.

Other considerations include knowing when or if you should prune. While pruning can help with overall vigor, if in doubt, prune your vines as soon as they finish flowering or don't prune them at all. This will reduce the risk of cutting off next year's blooms.

Keep in mind, too, that some perennial vines will take a number of years before they flower or fruit. Still, with the right growing conditions, perennial vines will reward you for years to come.

# FIVELEAF AKEBIA
*Akebia quinata*

This semi-evergreen vine provides quick cover, fragrant flowers, and—if you're lucky—interesting sausage-shaped fruits in autumn. The leaves, each one divided into five leaflets, are blue-green to green. But don't be fooled by its delicate appearance; this twiner can cover a wire fence in no time; it only needs a string or wire form to twine around. If left alone it will ramble and scramble as a groundcover. One word of caution: this vine can become a beast if you don't keep it trained by regular pruning.

**Size:** The leaves, 4 to 7 inches across, sometimes hide the small purple-brown flowers that appear in spring. The distinctive fruits are 2½ to 4 inches long. This vine will easily grow 20 to 40 feet or longer.

**Conditions:** Plant *Akebia* in full sun or part shade in well-drained soil, although it will tolerate a wide range of soil types.

**Zones:** 4 to 9

**Uses and Companions:** *Akebia* is ideal for covering an unsightly chainlink fence or as a green cover for a wood fence. Use it for screening or adding vertical interest and as a backdrop for evergreen and deciduous shrubs.

# CLIMBING ASTER
*Ampelaster carolinianus*

In late fall, October to November, Climbing Aster produces long, arching stems of purple daisies with yellow centers. This easy-to-grow native is a welcome sight when few other plants are in bloom.

**Size:** The long stems of Climbing Aster grow 6 to 10 feet tall and 2 to 2½ feet wide.

**Conditions:** Plant this vine in full sun or part shade in well-drained soil.

**Zones:** 6 to 9

**Uses and Companions:** This vine is perfect to train up through an open brick wall or wire fence. Let it ramble and scramble with spring-flowering shrubs, or grow it with Roses and Clematis to extend the season of bloom in your garden.

# CROSS VINE
*Bignonia capreolata*

In April it is a delight to walk in the woods and come upon the distinctive blooms of the cross vine. Tubular flowers, usually red or orange on the outside and yellow on the inside, stand out against semi-evergreen foliage. You may spot a few on the ground only to look up and discover the vine high in the treetops. Selections include 'Tangerine Beauty' with 2-inch-long flowers that are tangerine on the outside and yellow on the inside.

**Size:** Cross Vine grows 10 to 20 feet high or higher. This self-clinging vine attaches to trees and other surfaces with tendrils that won't damage plants or the surface it grows on.

**Conditions:** Cross Vine will grow in full sun or part shade. In shadier sites it will produce fewer flowers. Well-drained soil is ideal. Once it is established, Cross Vine will tolerate some drought.

**Zones:** 6 to 9

**Uses and Companions:** Use Cross Vine to cover a fence, wall, or pergola. Train it to grow up the trunk of a tree or over an evergreen shrub.

# CAROLINA JESSAMINE
*Gelsemium sempervirens*

A carefree native, Carolina Jessamine is also tough. In late winter to early spring it produces masses of fragrant, yellow flowers. Even when it is not in bloom, it offers handsome evergreen foliage. A word of caution: This plant is poisonous if ingested. The good news is that deer will avoid it.

**Size:** This vine will grow to 20 feet or higher.

**Conditions:** Plant this vine in full sun or part shade. In deep shade it will produce fewer blooms. Moist, well-drained soil is ideal. Once established, it will tolerate some drought. Prune it immediately after it flowers.

**Zones:** 7 to 9

**Uses and Companions:** This twiner is great for covering fences, tree stumps, or training up and over arbors; you will need to tie it up to get it started. It also makes an effective groundcover and can help control erosion. Grow it in combination with native Azaleas, Dogwoods, and Carolina Silverbell.

# CLIMBING HYDRANGEA

*Hydrangea anomala spp. petiolaris*

Although it will take several years to establish and several more before it flowers, the reward is worth the wait. Climbing Hydrangea has handsome foliage, beautiful white lacecap blooms, and often displays rich yellow foliage in autumn. As it matures it develops peeling, cinnamon-colored bark that adds to its beauty. It is truly a vine with four seasons of interest.

**Size:** The mature vine can reach 20 to 40 feet high or more with an equal spread. Start with the largest size plant you can find.

**Conditions:** Part sun or high-filtered shade is ideal. It likes moist, well-drained soil.

**Zones:** 4 to 7

**Uses and Companions:** Climbing Hydrangea attaches itself with tendrils called holdfasts to walls, fences, trees, and other surfaces. To get it started, you will need to tie it up. Train it up a mature tree, wall, or fence. Use it as a living trellis for annual vines while it becomes established. Plant it as a backdrop for Rhododendrons, Dogwoods, or Azaleas.

# VARIEGATED KADSURA

*Kadsura japonica*

This elegant vine offers glossy dark green foliage that turns shades of red and purple in the autumn. Variegated selections are appealing for year-round interest in the garden.

**Size:** A twining vine, Variegated Kadsura grows 8 to 15 feet tall, covering a structure in no time.

**Conditions:** Plant this evergreen to semi-evergreen vine in full sun to part shade in moist, well-drained soil. A male and female plant are required if you want the red fruits.

**Zones:** 7b to 11

**Uses and Companions:** Use this vine to cover an arbor, fence, or train up a wall. Use it as a backdrop for deciduous shrubs.

This vigorous vine has shiny, dark green, compound leaves (many smaller leaves make up the leaf) and is much better behaved in the garden than its cousin *Wisteria sinensis*. In late summer clusters of fragrant, magenta flowers appear on top of the foliage; the flowers have an earthy scent. A fast grower, it quickly covers an arbor or fence.

# EVERGREEN WISTERIA
*Milletia reticulata*

**Size:** A twining vine, Evergreen Wisteria can reach 20 feet tall.

**Conditions:** Plant this vine in full sun in well-drained soil.

**Zones:** 7 to 10

**Uses and Companions:** Train over an arbor, pergola, or wall. Combine it with an annual vine like Moon Vine.

# VIRGINIA CREEPER
*Parthenocissus quinquefolia*

A tough, low-maintenance vine, Virginia Creeper also makes a beautiful living wallpaper. The handsome foliage is made of leaves divided into five leaflets. Bright green all summer long, it turns shades of red to burgundy in fall.

**Size:** Attaching itself with tendrils that have sucker discs, this vine grows 30 to 50 feet or more.

**Conditions:** An adaptable plant, Virginia Creeper will grow in sun or shade in various types of soil. Once established, it will tolerate some drought. It is also salt tolerant.

**Zones:** 3 to 9

**Uses and Companions:** Virginia Creeper makes an effective groundcover, wall cover, or rambling vine over a pile of rocks. Avoid training this vine on shingles or wood siding, as the tendrils can be difficult to remove.

# JASMINE
*Trachelospermum jasminoides*
'Madison'

A profusion of fragrant, white, star-shaped flowers in early spring and its glossy evergreen foliage are two reasons to grow this beauty. Fast growing and tough, Jasmine is a winner for every season.

**Size:** A twiner, it grows 20 to 40 feet high.

**Conditions:** Plant 'Madison' in full sun or part shade in any average garden soil as long as it is not wet.

**Zones:** 7 to 10

**Uses and Companions:** Let this vine ramble as a groundcover, or help it get started to climb an arbor, fence, or structure. Plant it in an area with native Azaleas and spring-blooming bulbs.

# NATIVE AMERICAN WISTERIA
*Wisteria frutescens* 'Amethyst Falls'

If you love the Asiatic wisterias but want a vine that is easier to control and train in the garden, 'Amethyst Falls' is a choice alternative with one exception—the blooms are not as fragrant as the Asiatic types. The good news is this native will often produce its rich purple-blue flowers (4- to 6-inch-long clusters) in the first year, and it blooms in April, about three weeks later than the Asiatic types (which can be damaged by late frosts).

**Size:** Growing quickly to 30 feet, it can be pruned anytime since it blooms on current-year growth.

**Conditions:** Plant this Wisteria in full sun or part shade in well-drained soil.

**Zones:** 5 to 9

**Uses and Companions:** 'Amethyst Falls' is easy to train as a standard in a pot, over an arbor, or as an espalier.

EVERGREEN
CLEMATIS

# MORE PERENNIAL
## VINES

*Actinidia arguta* — Hardy Kiwi

*Clematis armandii* — Evergreen Clematis

*Clematis × jackmanii* — Clematis

*Decumaria barbara* — Native Climbing Hydrangea

*Lonicera × heckrottii* — Gold Flame Honeysuckle

*Lonicera sempervirens* — Trumpet Honeysuckle

*Parthenocissus henryana* — Silvervein Creeper

*Parthenocissus tricuspidata* — Boston Ivy

*Smilax smallii* — Jackson Vine

*Schizophragma hydrangeoides* 'Moonlight' — Japanese Hydrangea Vine

# THE ESPALIER

The term "espalier" dates to the seventeenth century and was originally applied to the method used for training fruit trees in open ground, either as permanent features or in preparation for placing them against walls. This type of artistic pruning is perfect if you want to grow ornamental trees or shrubs in a small space. Plants are grown flat like vines and trained against a wall, fence, building, trellis, or on a set of fixed wires. There are many different methods and styles of espalier, both formal and informal. Some look like giant fans and others are more formal and geometric in shape; for example, a single trunk whose branches have been trained to resemble candelabra. Plants in containers can also be trained as espaliers. Fruit trees such as dwarf apples, pears, quince, and figs work well because they are productive and they offer ornamental fruit and/or flowers. Other good candidates for espalier include vines, roses, *Camellia*, *Cotoneaster*, *Viburnum*, *Forsythia*, Witchhazels, and *Magnolia grandiflora*.

In a small space, espaliered plants, trees, and shrubs can lift your garden up walls and fences, while providing you with the opportunity to show off your imagination. And in larger spaces, espaliers can create dramatic backgrounds and accents.

PEAR ESPALIER

# ROSES

Historically, roses have been popular as garden shrubs but also considered high-maintenance plants that require lots of special attention and a regular spray program of pesticides and fungicides. You will notice that I have not included any Hybrid Tea Roses on my list; this is intentional. While there are no doubt many that are garden worthy, the roses I list here should thrive with a minimum of care and offer gardeners an introduction to growing some of these lovely plants. I have included repeat bloomers, fragrant selections, and a few climbers, offering a range of types for different garden settings.

Still, all roses perform best in full sun (a minimum of 6 hours), with regular applications of water and fertilizer, in soil that is amended with organic matter. While their foliage may not look good, when August comes I prefer to limit the use of chemicals in my garden, and therefore select roses that will thrive without spraying.

If, after growing some of these roses, you become enamored of the group and want to try growing Hybrid Teas, refer to the American Rose Society for recommended varieties. You may also check with a rose grower in the South, like The Antique Rose Emporium in Texas or Roses Unlimited in Laurens, South Carolina. Both of these firms offer a wealth of information about growing roses in the South.

## LADY BANKS' ROSE
*Rosa banksiae 'Lutea'*

Lady Banks' Rose has thrived in southern gardens for generations. Although the small, double, soft-yellow flowers look delicate, don't be fooled. This climbing rose is vigorous and tough. Make sure that you provide a substantial arbor or support for it to grow on, as it can get quite large over time. Handsome evergreen foliage, no thorns, and a profusion of flowers in early spring are all reasons to grow this carefree beauty. There is also a white-flowered form but it has some thorns.

**Size:** This rose can easily reach 20 feet high and wide or more.

**Conditions:** Lady Banks' Rose blooms in early spring before many other roses. Plant it in full sun in a well-drained soil. Prune immediately after it flowers, as blooms occur on two- and three-year-old wood.

**Zones:** 7 to 9

**Uses and Companions:** Train Lady Banks' Rose up over an arbor or pergola. Use it as a planting on a bank to help control erosion. Underplant it with early spring bulbs like the Daffodil, *Narcissus* 'Tete-a-Tete'.

## CAREFREE BEAUTY ROSE
*Rosa Carefree Beauty*™

Double pink, slightly fragrant flowers from summer to frost inspire gardeners to grow Carefree Beauty Rose. Add to this its disease resistance and compact rounded habit, and you have a great rose for the South.

**Size:** Carefree Beauty Rose grows 3 to 4 feet tall and wide.

**Conditions:** Plant this rose in full sun in well-drained soil. Grow it in a pot or in the ground.

**Zones:** 4 to 9

**Uses and Companions:** Incorporate Carefree Beauty™ into your perennial border, or plant it in combination with other Roses. It blooms from summer until frost. Underplant it with Hardy Geraniums, like *Geranium* 'Rozanne' with its 1-inch, violet-blue flowers.

# BUTTERFLY ROSE

*Rosa chinensis* 'Mutabilis'

This China rose is a choice plant for Southern gardens. Its fragrant and delicate single blooms are coral in bud, open to yellow, and then fade to pink and then crimson. The effect is as if a host of butterflies has landed on the plant. The new foliage and stems start out dark red, adding to the beauty of Butterfly Rose. Once established, this rose demands very little with its resistance to blackspot and powdery mildew, but offers blooms from spring until frost.

**Size:** Butterfly Rose grows 4 to 6 feet tall and 3 feet wide.

**Conditions:** Plant Butterfly Rose in full sun or part shade in well-drained soil. Prune only to remove dead wood or keep the plant in bounds.

**Zones:** 6 to 9

**Uses and Companions:** Plant Butterfly Rose with other roses or in the mixed border with shrubs and perennials. Underplant with *Veronica* 'Georgia Blue' or Geranium 'Rozanne' for a contrasting groundcover.

# KNOCK OUT® ROSE

*Rosa* Knock Out®

Clusters of 3½-inch, delicately tea-scented, cherry-red to scarlet blooms fill the landscape with color from early spring until fall. On top of that, this tough rose will tolerate drought once established. It is also resistant to pest and disease problems like blackspot and Japanese beetles. When it finally finishes blooming in late fall, the autumn foliage turns shades of deep red to violet and it produces orange-red hips.

**Size:** This is a shrub rose that grows 3 feet tall and wide, or larger.

**Conditions:** Plant Knock Out® Rose in full sun or part shade. Well-drained soil that is rich in organic matter is ideal. Unlike many repeat bloomers, it does not require deadheading or spraying with chemicals to keep foliage looking good. Prune only if plants get too large for its location.

**Zones:** 4 to 9

**Uses and Companions:** Knock Out® Rose is ideal for an informal hedge, backdrop, or screen. Add it to the perennial border for bright color all summer. Use it as a mass planting to line your driveway.

# POLYANTHA ROSE
*Rosa 'Marie Pavie'*

In spring, masses of delightfully fragrant blooms cover this small shrub rose. Pink buds open to reveal semi-double white flowers. After a major flush of flowers in spring, it blooms sporadically from summer until fall. Maturing at 4 feet by 4 feet, 'Marie Pavie' is a manageable size in the ground or in decorative containers. This old-fashioned Polyantha Rose has dark green foliage, very few thorns, and it is disease resistant.

**Size:** *Rosa* 'Marie Pavie' grows 3 to 4 feet tall and 2 feet wide.

**Conditions:** Plant this rose in full sun in moist, well-drained soil. Apply a 3- to 4-inch layer of mulch (organic matter) after it is planted.

**Zones:** 5 to 9

**Uses and Companions:** Plant 'Marie Pavie' in a large pot or in the mixed border with herbaceous plants like Foxglove, Iris, Hardy Geraniums, and Phlox. Use it for a low, formal hedge. A sport, or mutation, of this rose led to the introduction of *Rosa* 'Marie Daly', noted as an Earthkind rose; its pink buds open to fragrant, double-pink flowers.

# NEW DAWN CLIMBING ROSE
*Rosa 'New Dawn'*

A vigorous climber with glossy green foliage, 'New Dawn' blooms heavily with large pale pink, fragrant blooms in spring and then on and off from summer into fall. I grew this rose for years, trained along a picket fence, and it always looked good even when it wasn't blooming. It is tough and disease resistant to both blackspot and powdery mildew. I don't recommend letting annual vines grow up and through it since the rose's thorns make it hard to remove the vines once they finish blooming.

**Size:** 'New Dawn' grows 12 to 18 feet tall by 8 feet wide. Provide it with a strong support to grow on as mature plants are heavy.

**Conditions:** Plant 'New Dawn' in full sun or part shade in moist, well-drained, average soil. Once established, it will tolerate drought. Prune immediately after it flowers as most of the blooms form on old wood.

**Zones:** 5 to 9

**Uses and Companions:** This rambler is great for training up an arbor, pergola, or over a fence. Combine it with Foxgloves and Iris, or perennial vines like *Clematis* with its blue flowers for a contrast.

This has been a favorite of mine for years. With its small, pointed apricot buds that open to pale pink, pom-pom type flowers with a powerful scent, this rose never disappoints. It blooms in spring, on and off in summer, and in the fall as well. The apple-green foliage starts out with red tints and looks good most of the summer, reviving itself in the autumn. Its small stature, disease resistance, and few thorns are further reasons I like this charming rose.

# POLYANTHA ROSE
*Rosa* 'Perle d'Or'

**Size:** *Rosa* 'Perle d'Or' grows 4 to 6 feet tall by 4 feet wide.

**Conditions:** Plant 'Perle d'Or' in full sun or part shade in well-drained soil. Mulch with 3 to 4 inches of organic matter. Prune only to shape the plant and remove dead wood.

**Zones:** 6 to 9

**Uses and Companions:** Plant this rose in the mixed border with perennials like *Geranium* 'Rozanne', *Veronica* 'Georgia Blue', Siberian Iris, and fall-blooming Asters. It also makes a great container plant.

# RED CASCADE CLIMBING ROSE
*Rosa* 'Red Cascade'

This slightly fragrant, miniature rambling rose offers scarlet-red flowers during spring, summer, and fall. Although individual blooms are only 1 inch across, when this rose is in bloom, it puts on quite a show. It can be grown as a groundcover or trained up through a shrub or on a trellis. Unlike many roses it doesn't seem bothered by serious pest or disease problems. Once established, it is also drought tolerant.

**Size:** It can grow 8 to 10 feet tall or taller with a spread of 6 to 8 feet.

**Conditions:** Plant this miniature in full sun (a minimum of 6 hours) and well-drained soil. Limit pruning to removing dead wood in early spring or shaping plant. Topdress with 1 inch of organic material in early spring and fall.

**Zones:** 5 to 9

**Uses and Companions:** Red Cascade Climbing Rose looks great when paired with conifers like *Chameacyparis pisifera* 'Filifera Aurea' with golden variegated foliage. It puts on a show when trained to grow up a white trellis. Combine it with Iris, Phlox, and Baptisia.

# SHRUB ROSE
*Rosa rugosa*

If carefree is your idea of growing roses, then this is the rose for you. This distinctive rose has wrinkled glossy green leaves and intensely fragrant flowers. Depending on the selection, colors range from white to yellow, pink, and purple and are single, semi-double, or double. This rose blooms in early summer and then on and off until frost. The orange-to-red rose hips appear in summer and put on a beautiful display well into autumn. The only requirement of this vigorous rose is well-drained soil. It tolerates salt spray, some drought, and will even grow in pure sand.

**Size:** The Shrub Rose grows 4 to 6 feet high and 4 to 6 feet wide, or larger.

**Conditions:** Full sun or part shade is needed. Prune in early spring only to remove dead wood or to keep the plant from getting too large. Topdress with 1 inch of organic material in early spring and fall.

**Zones:** 2 to 9

**Uses and Companions:** Shrub Rose makes a great hedge or barrier due to its many thorns. Cultivars with high disease resistance include 'Albo-plena' with double white flowers; 'Belle Poitevine' with semi-double, mauve-pink flowers; and 'Fru Dagmar Hastrup' with intensely fragrant, silver-pink blooms.

# POLYANTHA ROSE
*Rosa 'The Fairy'*

Not all roses are created equal, and 'The Fairy' is no exception. What this charmer offers is a compact, disease-resistant shrub with masses of double pink blooms from late spring until frost. Longlasting as a cut flower, there can be up to 25 petals per flower. The only drawback is its lack of fragrance, but the profusion of flowers more than makes up for this.

**Size:** *Rosa* 'The Fairy' grows 2 to 3 feet high and 2 to 3 feet wide.

**Conditions:** Plant in full sun or light shade in moist, well-drained soil. Prune in early spring to remove dead wood. Topdress with 1 inch of compost in early spring and fall.

**Zones:** 5 to 9

**Uses and Companions:** This shrub is great for growing in containers or in the ground. Plant it at the front of the border with perennials like Iris, Rosemary, and Lavender, and let it arch over the edge. It can also be used as a groundcover.

'BELINDA'S DREAM'

# MORE ROSES

*Rosa* 'Belinda's Dream' — *Shrub Rose (Earthkind Rose)*

*Rosa* 'Buff Beauty' — *Hybrid Musk Rose*

*Rosa* 'Cecile Brunner' — *Polyantha Rose*

*Rosa* 'Climbing Pinkie' — *Polyantha Rose*

*Rosa* 'Marjorie Fair' — *Hybrid Musk Rose*

*Rosa* 'Mrs. B. R. Cant' — *Shrub Rose*

*Rosa* 'Old Blush' — *China Rose*

*Rosa* 'Pink Pet' — *Polyantha Rose*

*Rosa* 'Souvenir de la Malmasion' — *Shrub Rose*

*Rosa* 'Zephrine Drouhin' — *Cimbing Bourbon Rose*

# ROSE RUSTLIN'

Are you a potential rose rustler? Have you discovered an old Rose thriving along the roadside or in a cemetery despite harsh conditions and no special care? Did you get permission to take cuttings and propagate this treasure? If you answer yes to these questions, then you are on your way to becoming a rose rustler. The Texas Rose Rustlers have a goal to ensure that Old Garden Roses are grown and promoted in gardens everywhere. Roses have long been popular for their beautiful and fragrant flowers, but many of the modern types (like Hybrid Teas and Grandifloras) demand spraying on a regular basis. Old Roses, rescued from obscurity by the enthusiastic Rustlers, are grown on their own roots. In general, they tend to be hardier and more resilient than grafted Roses, making them ideal for many gardens.

Below you'll find some essential tools for collecting Old Roses.

## ROSE RUSTLER KIT

**Sharp pruners**—to take cuttings

**Gallon-sized zip-lock bags**—for cuttings

**Paper towels**—to wrap around cuttings to keep them moist

**Water**—to moisten the paper towels for transport

**A pencil**—to take notes and write labels

**3- by 5-inch index cards**—to make labels for individual plants

**A notebook to keep records in**—where and when you discovered a rose

**A camera that is easy to operate**—to photograph the plants you find

Discover more about rose rustling at www.texasroserustlers.com.

ROSE ON COUNTRY FENCE

ROSA
'CECILE BRUNNER'

# SPRING FLOWERING
## SHRUBS AND TREES

Spring is a time when we get excited about gardening and often purchase plants on impulse. But the most successful gardens are based on the concept of using the right plant for the right place and combining them with other plants that like a similar environment. Hiring a professional to help you develop an overall plan is a good place to start. Another idea is to keep a list of what you like and where you plan to use it in your landscape, as well as what its companions will be; this way, you will be ready when you find that perfect plant.

Beyond Azaleas there are many shrubs and trees that thrive in our Southern gardens, including natives and exotics. Planting shrubs and trees in combination with one another, like native Azaleas with Dogwoods and Redbuds, is one approach.

If you need to prune your spring flowering shrubs and trees, the best time is immediately after they finish flowering. However, certain plants, like the Oakleaf Hydrangea, don't require pruning unless the plant is getting too big or there is dead wood that needs to be removed.

Knowing what shrubs and trees look like when they are not in bloom is useful information and may help you decide where to locate them in the garden.

# RED BUCKEYE
*Aesculus pavia*

Easy to grow, Red Buckeye offers handsome foliage, beautiful blooms, and disease resistance. A native, it makes a choice large shrub or small tree at the edge of the woodland, or as a specimen in the mixed border. When it blooms in spring, it puts on a show when the 4- to 8-inch-long flowers—or the rare yellow selection—stand up in spikes above the classic Buckeye leaves, divided like a fan into 5 to 7 leaflets.

**Size:** Red Buckeye matures at 10 to 20 feet high and wide, or wider.

**Conditions:** Full sun or part shade is ideal, but this adaptable native will also tolerate a good bit of shade. Moist, well-drained soil is best.

**Zones:** 4 to 8

**Uses and Companions:** Plant Red Buckeye in natural settings, and combine it with plants like native Azaleas and the groundcover *Pachysandra procumbens*, Dwarf Crested Iris, and Woodland Phlox like *Phlox divaricata*. Plant it in a shrub border against a backdrop of evergreens such as American Holly or Tea Olive, *Osmanthus fragrans*.

# DOWNY SERVICEBERRY
*Amelanchier arborea*

As one of the first trees to bloom in spring, Serviceberry lights up the woodland with its white flowers. The foliage emerges a gray-green and turns dark green as the summer progresses. With the arrival of autumn, the leaves change to shades of yellow, red, and orange. In June the edible small fruits, ¼ to ⅓ inch in diameter, turn from green to red and then black when they're ripe. You'll have to be quick if you want to sample the fruits before the birds devour them.

**Size:** A large shrub or small tree, Downy Serviceberry grows 15 to 25 feet or higher with a spreading habit. Usually multi-stemmed, it can be trained to grow as a single trunk. Look for 'Autumn Brilliance', which is noted for its brilliant fall foliage.

**Conditions:** Full sun or part shade and moist, well-drained soil are ideal.

**Zones:** 4 to 9

**Uses and Companions:** Downy Serviceberry is a native that works well where a small tree is needed as a specimen or focal point. Plant it in the shrub border in combination with other natives like Sweetshrub, *Calycanthus floridus*, or Summersweet, *Clethra alnifolia*.

# SWEETSHRUB
*Calycanthus floridus*

Just one flower can fill the air with its sweet scent, a combination of strawberries and bananas. Sweetshrub has pungent, dark maroon blooms (2 inches across) that perfume the landscape especially on warm, sunny days. The leaves and stems are also fragrant when crushed. In autumn the foliage turns from dark green to bright yellow. Some selections are more fragrant than others so try to purchase this plant when it's in bloom.

**Size:** Sweetshrub grows 6 to 9 feet tall with a 6- to 12-foot spread.

**Conditions:** Grow Sweetshrub in full sun or part shade. Moist, well-drained soil is ideal, but it will tolerate a range of soil types. This is a carefree shrub with no serious pest or disease problems.

**Zones:** 4 to 9

**Uses and Companions:** Plant Sweetshrub in the shrub border or at the edge of a woodland. Site it near paths so it will be easy to experience the fragrant blooms. Combine it with native Azaleas and evergreen shrubs.

# REDBUD
*Cercis canadensis*

Redbud is one of the first trees to bloom in spring; its flowers start out reddish purple in bud and open to a rosy pink. Quickly following are the distinctive, dark green heart-shaped leaves. Autumn color varies from yellow-green to pure yellow. For more colorful leaves, those of 'Forest Pansy' emerge a red-purple that fades somewhat as the season progresses.

**Size:** Redbud can grow 20 to 30 feet tall with a spread of 25 to 35 feet. Often wider than they are long, the leaves are 3 to 5 inches in length.

**Conditions:** Redbud prefers full sun or part shade and moist, well-drained soil, but it will adapt to a range of soil types.

**Zones:** 4 to 9

**Uses and Companions:** Redbud is a choice tree for native landscapes or woodland gardens. Combine it with Dogwoods, native Azaleas, and evergreen Hollies.

# FRINGE TREE OR CHINESE FRINGE TREE

*Chionanthus retusus*

Chinese Fringe Tree is stunning when it blooms in spring. Masses of delicate, fleecy white blooms cover the tree. Borne on upright panicles, 2 to 3 inches high and 2 to 4 inches wide, the blooms dance in the breeze. The green leaves are both leathery and lustrous. With its beautiful blooms, blue-black fruits (only on female trees), rich foliage, and handsome gray-brown bark, this small tree gives four seasons of interest.

**Size:** Chinese Fringe Tree grows 15 to 25 feet high or higher with a rounded, spreading habit.

**Conditions:** Plant Chinese Fringe Tree in full sun or part shade in moist, well-drained soil.

**Zones:** 5 to 8

**Uses and Companions:** Chinese Fringe Tree makes a choice specimen either as a large shrub or small tree. Resistant to pest and disease problems, it will also grow in urban sites. You can grow this tree in combination with the native Fringe Tree, *Chionanthus virginicus*, which blooms 2 to 3 weeks later and with slightly larger blooms than *C. retusus*.

# FLOWERING DOGWOOD

*Cornus florida*

This small- to medium-sized native tree offers beauty in every season. In spring it is covered with white blooms (the showy parts are really bracts), and in summer the bright green foliage provides welcome shade. In fall the leaves turn shades of red, orange, and purple before they drop and the bright red fruits are highlighted. In winter the bark and branches provide interesting texture and form. Look for improved selections like 'Cloud Nine' that are better adapted to tolerate heat.

**Size:** Flowering Dogwoods grow 20 to 40 feet tall with a spread equal to or greater than its height.

**Conditions:** Full sun or part shade is ideal. In full shade, Dogwoods will not flower well. Moist, fertile soil is best; Dogwoods are not known for being drought friendly.

**Zones:** 5 to 9

**Uses and Companions:** Flowering Dogwood is ideal as a small specimen tree or in a mass planting. Planting it against a backdrop of evergreens will highlight its winter silhouette. Combine it with native Azaleas and Redbuds.

# CAROLINA SILVERBELL

*Halesia tetraptera*

Few trees are as charming when they're in bloom as the Carolina Silverbell. Looking up into a tree that is covered with white, bell-shaped flowers is a delight in spring. The curious, four-winged seedpods that follow add another element of interest and persist long after the leaves have dropped. The bark, a combination of gray to brown to black, adds ornamental appeal too.

**Size:** This tree grows 30 to 40 feet tall with a spread of 20 to 35 feet.

**Conditions:** Carolina Silverbells perform best in full sun or part shade in moist, well-drained soil.

**Zones:** 4 to 8

**Uses and Companions:** Perfect for the edge of a woodland, this native tree looks especially good when the bells dance in the breeze. Combine Carolina Silverbell with Rhododendrons, native Azaleas, and perennials like *Phlox divaricata* and *Iris cristata*. Site it against a backdrop of evergreens to highlight its blooms.

# OAKLEAF HYDRANGEA

*Hydrangea quercifolia*

This adaptable native offers beautiful blooms, ornamental peeling bark, and attractive foliage that resemble large oak leaves. In late spring large, 4- to 12-inch panicles of erect white flowers put on a show that continues even as they fade. The bracts turn from white to pink to purplish-pink and then brown. In autumn the leaves take on varying shades of maroon, red, orange, and purple. As the plant matures the branches and trunk peel to expose inner layers of brown and cinnamon-colored bark.

**Size:** Oakleaf Hydrangeas grow 4 to 6 feet tall or taller with an equal spread.

**Conditions:** Full sun or part shade and moist, well-drained soil are ideal. Once established, this Hydrangea is drought friendly. Keep pruning to a minimum, and prune immediately after plants flower.

**Zones:** 5 to 9

**Uses and Companions:** For large bold texture in the landscape, Oakleaf Hydrangea is a winner. Site it at the back of the shrub border or in the woodland garden with Azaleas, Hostas, and Ferns. Selections include 'Pee Wee', a compact form that grows about 3 feet tall, and 'Snowflake' with double flowers.

For an elegant evergreen tree that thrives in heat and humidity, Southern Magnolia has long been a favorite. The thick, leathery leaves look good all year, and the fragrant white flowers, 8 to 12 inches in diameter, appear in late spring to early summer. The fruits, a cone-like structure that opens to expose bright red seeds in fall, are popular for use in holiday decorations. If space is a problem, look for more compact selections like 'Bracken's Brown Beauty' or the dwarf 'Little Gem', which should mature at 20 feet tall or less.

# SOUTHERN MAGNOLIA
*Magnolia grandiflora*

**Size:** Southern Magnolia matures at 60 to 80 feet in height with a spread of 30 to 50 feet; this tree needs lots of room.

**Conditions:** Full sun or part shade and moist, well-drained soil are ideal. Known for dropping lots of leaves, make sure you plant this tree where you won't be bothered by leaf litter.

**Zones:** 6 to 9

**Uses and Companions:** It's a great tree for screening, a hedge, or where a large evergreen specimen is needed. Use this as a backdrop to deciduous Magnolias or small, flowering trees like Dogwoods and Redbuds.

# FLORIDA AZALEA
*Rhododendron austrinum*

Stunning in bloom, this native offers fragrant flowers that range from yellow to cream, orange, and almost pure red. Like many other native Azaleas, it is also easy to grow and requires minimal maintenance. There are numerous selections, including 'Austrinum Gold' with bright gold, fragrant flowers. The beauty of native Azaleas is that once they finish blooming, the flowers dry up and disappear instead of clinging on like dead tissue paper the way many of the hybrid Azalea blooms do.

**Size:** Florida Azalea grows 8 to 10 feet high and at least half as wide, forming a loose, open shrub.

**Conditions:** This deciduous Azalea will tolerate full sun provided it has moist, well-drained soil. Otherwise, it performs best in light shade. Don't prune unless it gets too large for the spot where it's sited.

**Zones:** 7 to 9

**Uses and Companions:** Plant Florida Azalea with other native Azaleas for a sequence of bloom. Use them in the shrub border with a mixture of evergreen and deciduous plants. Underplant them with spring wildflowers like Phlox, Green and Gold, or Mayapple.

# MORE FLOWERING SHRUBS AND TREES
CAMELLIA

FOR SPRING

*Camellia japonica* cultivars — *Camellia*

*Deutiza gracilis* — *Slender Deutzia*

*Kerria japonica* — *Japanese Kerria*

*Magnolia* 'Butterflies' — *Butterflies Magnolia*

*Malus* 'Sugar Tyme' — *Crabapple*

*Prunus × yedoensis* — *Yoshino Cherry*

*Spiraea thunbergii* 'Ogon' — *Ogon Spiraea*

*Styrax japonicus* — *Japanese Snowbell*

*Styrax obassia* — *Fragrant Snowbell*

*Viburnum macrocephalum* — *Chinese Snowball*

# WEEPING TREES

Weeping trees add architectural interest to the garden, especially in winter, when deciduous trees take the stage with their interesting branching and dramatic silhouettes. Conifers, like *Cedrus atlantica glauca* 'Pendula', provide year-round interest and can be trained to grow over a wall or rock.

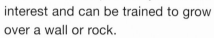
ATLAS CEDAR

Although there are weeping forms of large trees like the European Beech, *Fagus sylvatica*, and the Weeping Linden, *Tilia petiolaris*, there are also weeping trees that do not take up as much space as some other species.

As gardens get smaller, weeping trees and shrubs offer a way to add interest to the landscape, even when space is limited. Some weepers are also good candidates for containers. There are both deciduous and evergreen types to provide year-round interest in the garden and brighten the winter scene.

While some weeping plants occur naturally in the wild, the next generation of plants must be propagated from cuttings or grafted to carry on the traits of the parents. With weepers that have a graft close to the ground, it is recommended that you stake the central leader on the main trunk when it is young. Otherwise, you may end up with a tree that cascades to form a mound rather than an elegant weeper.

## Some Good Weepers

*Cercidiphyllum magnificum* 'Pendulum'—Weeping Katsura
*Cercis canadensis* 'Traveller'—Weeping Redbud
*Cercis canadensis* 'Lavender Twist' also known as 'Covey'—Weeping Redbud
*Prunus mume* 'Pendula'—Weeping Flowering Apricot

VIBURNUM MACROCEPHALUM
CHINESE SNOWBALL

# SUMMER FLOWERING
## SHRUBS AND TREES

Summer arrives quickly in the South, and many shrubs and trees are standing by waiting for their moment to shine. As challenging as it can be to garden during periods of drought and extreme heat, there are still rewards. Once they are established in the garden, shrubs like Butterfly Bush and Rose of Sharon bloom all summer and into fall and require only minimal care. Another carefree beauty is Tardiva Hydrangea. It produces large, upright panicles of white flowers in August when few other shrubs are in bloom. Some, like the dwarf selection of Pomegranate called 'State Fair', seem to be constantly in bloom or fruit.

Take time to learn about individual plants and whether they bloom on new wood or second-year wood. This will help you decide when and how much to prune your shrubs. For example, Butterfly Bush and Tardiva Hydrangea bloom on new growth, so you can prune them back hard every year in early spring just before new growth starts without fear of cutting off potential blooms.

Incorporate these summer bloomers into the herbaceous or shrub border to complement perennials, annuals, and bulbs, or grow them alone to show off their individual beauty.

# BOTTLEBRUSH BUCKEYE
*Aesculus parviflora*

This native beauty offers interest during spring, summer, and fall. In early spring dark green palmate leaves with 5 to 7 leaflets, ranging from 3 to 8 inches long, emerge on this large shrub or small tree. The striking white flowers sit atop the foliage and look like bottlebrushes. The blooms appear in summer from June to July on 8- to 12-inch-tall spikes. The show continues in autumn when the leaves turn golden yellow. This is a great understory shrub or specimen.

**Size:** Bottlebrush Buckeye grows 12 to 15 feet tall and 12 to 15 feet wide.

**Conditions:** Plant this large shrub or small tree in part shade or full sun in moist, well-drained soil. Bottlebrush Buckeye spreads by suckers, and although it does not require regular pruning, you can prune back almost to the ground if it needs to be rejuvenated.

**Zones:** 4 to 9

**Uses and Companions:** Plant Bottlebrush Buckeye at the edge of a woodland or under the shade of large trees like Oaks and Pines. Underplant it with a groundcover like *Phlox stolonifera* or Lenten Roses, *Helleborus orientalis*. It makes an effective specimen in the middle of the lawn.

# BUTTERFLY BUSH
*Buddleia davidii* and cultivars

This easy-to-grow, nonstop summer bloomer produces fragrant and colorful spikes of flowers that attract butterflies and humans alike. Butterfly Bush is a large, arching shrub that looks good planted in groups or as a specimen in the middle of the flower garden. Depending on the selection, the foliage is gray-green to dark green above and fuzzy beneath. There are numerous flower color selections including deep purple, lavender, white, yellow, deep red, and pink. Flowers occur on new wood, so prune in early spring before new growth starts for the best blooming.

**Size:** This fast-growing shrub can easily reach 10 to 15 feet high and half as wide in one season. There are dwarf selections like 'Nanho Purple'.

**Conditions:** Full sun and moist, well-drained soil are ideal. Fertilize with a 10-10-10 in early spring and again in summer. If plants get too large you can prune them back during the growing season. Deadhead spent flowers to encourage more blooms.

**Zones:** 5 to 9

**Uses and Companions:** Plant Butterfly Bush in the perennial or shrub border for months of color. Combine it with Lantana, Daylilies, and Ornamental Grasses.

# GARDENIA
*Gardenia jasminoides*

Certain scents bring back strong memories for me, and Gardenia tops the list. Its powerfully sweet flowers remind me of my grandmother and the Gardenias she grew in her backyard. This old-fashioned plant still holds up in today's garden with its intoxicating blooms and glossy green foliage. There are a number of cultivars available including 'Radicans', a prostrate creeping form with 1-inch-diameter blooms that makes a good edger but is not as reliably hardy as selections like 'Klein's Hardy', which grows 2 to 3 feet tall and has single blooms.

**Size:** Depending on the species or cultivar, Gardenias grow 2 to 6 feet tall or taller and 2 to 4 feet wide. 'August Beauty', which has double blooms, grows 4 to 6 feet high.

**Conditions:** Plant gardenias in full sun or part shade in an acidic soil that is moist and well drained.

**Zones:** 7 to 10

**Uses and Companions:** Gardenias can make effective foundation plants or specimens. Use them in formal or informal gardens for hedges, edging, or as a backdrop in the perennial border. Combine them with shrubs like Butterfly Bush and summer bloomers like Lantana and Heucheras.

# ROSE OF SHARON
*Hibiscus syriacus*

This large shrub or small tree thrives in the heat and blooms for weeks, even during the "dog days" of summer when many other plants look tired and wilted. Once established, Rose of Sharon is also drought friendly. There are numerous selections with single, semi-double, and double blooms. Selections that are noted for producing larger flowers and few or no seedpods include 'Diana', with pure white blooms, and 'Helene', which has white flowers with a deep red eye. 'Bluebird' has single blue blooms with a deep red eye.

**Size:** This large shrub can easily reach 10 to 15 feet tall and 2 to 3 feet across. The flowers are 2½ to 3 inches across.

**Conditions:** Full sun or part shade and average, well-drained soil is all this plant needs. Cut back the previous season's growth in late winter to encourage bigger blooms.

**Zones:** 3 to 8

**Uses and Companions:** Plant this shrub at the back of the border or on the sunny edge of a natural area. Plant Rose of Sharon against evergreens, as it is late to leaf out in spring. Combine it with Daylilies, Daisies, and Black-eyed Susans.

# TARDIVA HYDRANGEA
*Hydrangea paniculata* 'Tardiva'

This selection of *Hydrangea paniculata* is one of my favorites for its late-season bloom (August or later) and for how easy it is to grow. Because it blooms on new wood, you don't have to worry about pruning off next year's blooms. Large, white cone-shaped panicles of flowers grace this shrub and are attractive for months. Even after the blooms fade, and 'Tardiva' drops its leaves, the dried flower-heads persist well into winter, providing textural interest in the landscape.

**Size:** This Hydrangea can grow 6 to 8 feet tall and 2 to 3 feet across. The flowerheads are 12 to 18 inches tall.

**Conditions:** This hydrangea thrives in full sun and moist, well-drained soil. Prune it back in early spring before new growth starts.

**Zones:** 3 to 8

**Uses and Companions:** Plant 'Tardiva' in the perennial or shrub border against a backdrop of evergreens. Use it as a specimen or focal point for the summer garden. Underplant it with white flowers like *Erigeron karvinskianus* (Fleabane) or the annual *Euphorbia* 'Diamond Frost'.

# ST. JOHNSWORT
*Hypericum calcyinum, H. frondosum,*
*H.* 'Hidcote'

Creeping St. Johnswort, *H. calycinum,* makes an effective low groundcover, to about 1 foot high, at the front of the border or on a slope or bank where it can help control erosion. The bright yellow flowers are 3 inches across and bloom for weeks in summer. There are also shrubby types like *H. frondosum*, with

mostly evergreen blue-green foliage and bright yellow flowers that grows 3 feet tall or taller and wide. *Hypericum* 'Hidcote', which is also semi-evergreen, grows 4 feet tall and produces masses of yellow flowers all summer.

**Size:** Depending on the selection, St. Johnswort grows 1 to 5 feet tall and half as wide.

**Conditions:** Full sun or part shade and moist soil provide the ideal conditions. For groundcover types, mow or prune back the tops every two to three years to keep plants vigorous. With shrubby types prune out dead wood as soon as plants finish blooming.

**Zones:** 5 to 8

**Uses and Companions:** St. Johnswort makes an effective groundcover, mass planting on a bank or slope, informal hedge, or a border. Combine it with other summer bloomers like Daylilies, Black-eyed Susans, and Coneflowers.

# GOLDEN RAINTREE

*Koelreuteria paniculata*

This adaptable, drought-friendly tree is a most welcome sight in summer when it is covered with panicles of bright yellow flowers. The divided, dark green leaves look good all summer. The seed capsules that appear after it flowers look like miniature paper Japanese lanterns and change from green to yellow and then brown. Even showier are the pink-to-rose seed capsules of the related species *Koelreuteria bipinnata*, also called the Bougainvillea Golden Raintree.

**Size:** Golden Raintree grows 30 to 40 feet tall and almost as wide.

**Conditions:** This tree performs best in full sun but will tolerate a wide range of soil types. It also tolerates drought and air pollution. If you need to prune it to shape the tree, do it in late winter.

**Zones:** 5 to 9

**Uses and Companions:** Plant Golden Raintree as a street tree, specimen, or small patio tree where space is an issue. Combine it with ornamental grasses and other summer bloomers like Butterfly Ginger, *Hedychium coronarium*.

# POMEGRANATE

*Punica granatum*

Drought tolerant and easy to grow, Pomegranates give a lot and ask for little in return. They have handsome foliage, beautiful orange tubular flowers, and red-orange fruits. The selection 'Wonderful' is self-fruiting and doesn't require a pollinator. The green foliage with orange-red markings also turns yellow in fall when the fruits appear. If you don't have lots of space and want a hardy type, try the dwarf selection *Punica granatum* 'State Fair', which throughout summer and fall always seems to have fruit and flowers.

**Size:** The species grow 20 feet tall and 5 feet wide. The selection 'State Fair' has silver-dollar-sized fruits and grows to a height of 4 to 5 feet.

**Conditions:** Full sun is best, but Pomegranates will grow in a wide range of soil types. Prune them in late winter to shape plants if needed.

**Zones:** 7 to 10

**Uses and Companions:** Plant Pomegranate as a specimen or part of a mixed shrub border. Dwarf Pomegranate is also a candidate for the rock garden. Combine it with hardy Salvias and spring bulbs.

## PLUMLEAF AZALEA
*Rhododendron prunifolium*

Native Azaleas are always a good choice for our Southern gardens, and the Plumleaf Azalea is a gem. In July to August brilliant red-orange to red flowers cover this deciduous shrub. A butterfly magnet, it lights up the garden at a time when little else is in bloom. Once it finishes flowering, the dead flowers disappear, and you have a green shrub for the rest of the growing season.

**Size:** This native grows 8 to 10 feet tall or taller. The funnel-shaped flowers are ¾ to 1 inch long.

**Conditions:** In its native habitat, this Azalea is found growing along streams, but in the garden it will thrive in full sun or part shade and moist, well-drained soil. Limit pruning to removing dead wood. It is pest and disease resistant.

**Zones:** 5 to 9

**Uses and Companions:** Combine Plumleaf Azalea with other Native Azaleas for a long season of bloom. Plant them against an evergreen backdrop of hemlocks or *Cryptomeria japonica*. Plant Plumleaf Azalea near Cardinal Flower.

## CHASTE TREE
*Vitex agnus-castus*

Fragrant foliage and spikes of lilac-blue flowers that last for weeks in summer are the rewards from growing this large shrub or multi-trunked small tree. More reasons to grow Chaste Tree: it is fast growing, thrives in heat, tolerates a range of soil types, and is pest resistant. The gray-green, aromatic leaves are divided like fans with 5 to 7 leaflets. There are selections with white, pink, and deep blue flowers.

**Size:** Depending on where it grows, Chaste Tree can reach 10 to 25 feet tall and half as wide. The 6- to 12-inch spikes of flowers sit atop leaves that are 2 to 6 inches long.

**Conditions:** Full sun and well-drained soil are best for this summer bloomer. It blooms on current season's growth, so prune in late winter before new growth starts.

**Zones:** 6 to 8

**Uses and Companions:** Plant a Chaste Tree at the back of the perennial border, in the mixed shrub border, or as a specimen. Combine it with Salvias, Ornamental Grasses, and late-blooming varieties of Daylilies.

# MORE FLOWERING SHRUBS AND TREES
## FOR SUMMER

JAPANESE PAGODA TREE

*Clerodendrum trichotomum* — Harlequin Glorybower

*Cornus kousa* — Kousa Dogwood

*Franklinia alatamaha* — Franklin Tree

*Hydrangea paniculata* 'Limelight' — Hardy Hydrangea

*Koelreuteria bipinnata* — Chinese Flame Tree

*Lagerstroemia indica* cultivars — Crapemyrtle

*Magnolia virginiana* — Sweet Bay Magnolia

*Styphnolobium japonicum* — Japanese Pagoda Tree

*Stewartia monadelpha* — Tall Stewartia

*Stewartia pseudocamellia* — Japanese Stewartia

# THE LANDSCAPED CEMETERY

The first landscaped cemetery, Père Lachaise Cemetery, opened in Paris, France, in 1804 as a reaction to increasing urban populations and crowded church graveyards.

In the United States, rural garden cemeteries began to appear in the mid-1800s that incorporated architecture and design as well as plants of all types, both native and exotic. Before there were public parks, cemeteries were designed for public use. It was not uncommon to take a Sunday stroll or have a picnic at a cemetery. There are now garden cemeteries throughout the United States. Today, one of the most well known for its gardens and plants is Mt. Auburn near Boston, Massachusetts. It was founded in 1831, and its collection emcompasses over 175 acres and includes examples of different styles of landscapes, ranging from the Victorian to the contemporary. There are over five thousand trees representing 630 taxa.

In Atlanta, Georgia, Oakland Cemetery, established in 1850, is another shining example. Today it encompasses forty-eight acres and is named for its many Oaks, symbols for strength, longevity, and protection. There are many different types of garden styles represented and plants of all types for every season. The staff at Oakland Cemetery is working to establish an arboretum.

STEWARTIA PSEUDOCAMELLIA
JAPANESE STEWARTIA

Whether they hide unsightly areas, provide a hedge between neighbors, or divide different sections of the garden, shrubs for screening and hedging play an important role in the landscape. In some cases they take the place of a fence, and in others they offer an affordable solution to soften the effect of a utilitarian fence (chain-link, wire, or wood). There are deciduous varieties of shrubs that are effective but evergreens are more suited for screening. In some cases a combination of evergreen and deciduous shrubs makes for an effective hedge or screen.

Some evergreens, like Camellias, offer beautiful foliage and flowers; others, like Tea Olive, have handsome leaves but the flowers are insignificant (although their perfume is delightful). American Holly, *Ilex opaca*, offers colorful fruits and glossy foliage. While the focus of this chapter is on shrubs, I have included one deciduous tree, European Hornbeam, *Carpinus betulus*, because certain forms work so well for hedging and screening.

The style of the hedge or screen will in large part be determined by how it is pruned. Sheared shrubs tend to have a more formal look, while those that are left to grow in a more natural form are better suited for informal settings.

# GLOSSY ABELIA
*Abelia × grandiflora*

This old-fashioned, easy-care shrub is a favorite of butterflies, bees, and birds and looks good during every season. Fast growing and mostly evergreen, it is ideal for an informal screen or hedge. The shiny foliage is dark green in summer and then takes on tinges of bronze and red in fall. It flowers on new growth from late spring until frost, producing a profusion of funnel-shaped flowers that are white flushed with pink. Its open, twiggy habit and graceful arching branches provide the perfect backdrop.

**Size:** This shrub grows as wide as it does tall, from 3 to 6 feet or taller. The leaves are ½ to 1½ inches long and half as wide.

**Conditions:** Plant Glossy Abelia in full sun or part shade in moist, well-drained soil. Prune for rejuvenation and to enhance its natural open, arching habit in late winter or very early spring before new growth starts.

**Zones:** 5 to 8

**Uses and Companions:** Use Glossy Abelia as a backdrop in the perennial border. Combine it with other evergreens and conifers. It makes a good companion with roses like the Knock Out Roses.

# FLORIDA LEUCOTHOE
*Agarista populifloia*

Florida Leucothoe is a native evergreen shrub that thrives in the shade and offers a graceful screen. With tall, arching branches it forms a fountain of foliage, perfect for an evergreen backdrop with deciduous trees or shrubs or a screen between two different areas of the landscape. The glossy, 4-inch-long rich green leaves look good during every season on this multi-stemmed shrub.

**Size:** An arching shrub with glossy evergreen foliage, Florida Leucothoe grows 8 to 12 feet tall and 5 feet wide. It makes a good mass planting for screening or an evergreen backdrop.

**Conditions:** Plant it in part to full shade. This native does not like to dry out, so keep the soil moist. Prune in early spring by removing any dead or declining individual stems at the base of the plant.

**Zones:** 6 to 10

**Uses and Companions:** Plant Florida Leucothoe at the edge of the woodland garden for a screen. Combine it with Rhododendrons, native Azaleas, and other shade lovers like Ferns, Hostas, and Coralbells. Use it as an evergreen backdrop for trees like the Coralbark Maple with its showy red-orange trunk and twigs.

This evergreen shrub, which shows up in many old, established gardens in the South, is still popular today. The small, rich green leaves, dense twiggy habit, and slow growth make it well suited for formal hedges, edging, and topiary. The foliage has a distinct scent that some find offensive and others don't notice. The selection 'Suffruticosa' is often considered the true edging Boxwood or English Boxwood because of its slow growth habit. Boxwood responds well to pruning and shearing but left unpruned, they form billowy shapes.

## AMERICAN BOXWOOD
*Buxus sempervirens*

**Size:** While the species can grow 15 to 20 feet high and wide, there are many slow-growing selections that mature at half this size. Without pruning, certain selections grow 4 to 5 feet tall, but they can also maintained at just a few inches high.

**Conditions:** Plant American Boxwood in moist, well-drained soil in full sun or part shade. To keep plants healthy, prune out dead center branches to allow light to get to the center of plants and encourage new growth.

**Zones:** 5 to 8

**Uses and Companions:** Use American Boxwood for formal hedges, foundation plants, or screening. Combine them with perennials like Autumn Fern and Hostas for a contrast in texture.

## CAMELLIA
*Camellia japonica* and *Camellia sasanqua*

Camellias offer elegant evergreen foliage, beautiful blooms, and good looks during every season. By planting both types, you can have blooms from early fall through late spring. Depending on how they are pruned, they work equally well in both formal and informal gardens. *Camellia japonica* blossoms, double or single, can be as large as 5 inches across and as small as 2½ inches, while *Camellia sasanqua* foliage and flowers tend to be smaller.

**Size:** *C. japonica* cultivars range from 10 to 25 feet tall and wide, while *C. sasanqua* ranges from 1½ to 15 feet.

**Conditions:** Full sun or part shade and moist, well-drained soil that is rich in organic matter is ideal for Camellias. Prune immediately after flowering. While Camellias can tolerate full sun, protect them from the hot afternoon sun for best success.

**Zones:** 7 to 9

**Uses and Companions:** Camellias work well not only as evergreen screens and hedges, but they also make choice specimen plants. Combine them with spring-flowering shrubs.

PROVEN PLANTS

# EUROPEAN HORNBEAM

*Carpinus betulus*

Tough and adaptable, European Hornbeam is also elegant, in particular the selection 'Fastigiata', which makes a choice hedge, screen, or backdrop for other shrubs and trees. The dark green, serrated leaves look good all spring and summer and then in fall they change to yellow or yellowish green. In winter a silhouette of fanlike branching and fluted gray bark add to its appeal. Another selection to look for is 'Columnaris'.

**Size:** European Hornbeam usually matures to 40 to 60 feet in height by 30 to 40 feet wide, but this tree can also grow much taller.

**Conditions:** European Hornbeam will grow in a range of soil types but is happiest with moist, well-drained soil in full sun or part shade. It also tolerates a good deal of pruning.

**Zones:** 4 to 7

**Uses and Companions:** Train European Hornbeam as a formal hedge. Use it to divide different sections of the garden into "rooms." Plant it against a backdrop of evergreen hollies and conifers.

# AMERICAN HOLLY

*Ilex opaca*

This extremely hardy native has glossy evergreen foliage and colorful fruits all winter that appeal to birds and humans. There are selections with red, orange, and yellow fruits. The cut branches are also popular for holiday decorations. Varying in habit from upright and broad to narrow, or low and spreading, American Holly is a choice evergreen that serves myriad functions in the landscape. You need one male pollinator of the same species for a group of three females to ensure good fruit set with Hollies.

**Size:** American Holly grows 15 to 30 feet tall by 10 to 20 feet wide.

**Conditions:** Full sun or part shade and well-drained soil are ideal, but Hollies will tolerate a range of growing conditions.

**Zones:** 5 to 9

**Uses and Companions:** American Holly makes an effective formal or informal screen or hedge depending on how it is pruned. Pair it with deciduous Hollies for a spectacular effect in winter. Dogwoods and native Azaleas also make happy companions for American Holly.

With its handsome olive-green, anise-scented foliage, this native evergreen makes a choice screen or backdrop in the garden. Perfect for shady, damp locations, it will also adapt to sunnier sites. With its large leaves and upright habit, this Anise adds year-round color and texture to the garden. Crush the leaves to release the anise scent. Insignificant flowers are followed by curious, star-shaped seedpods.

## SMALL ANISE TREE
*Illicium parviflorum*

**Size:** Small Anise Tree grows 10 to 15 feet tall by 6 to 10 feet wide.

**Conditions:** Part to half shade and moist, well-drained soil are ideal. Anise will grow in full sun, but the leaves can become bleached looking, especially in dry soils. Prune this shrub by hand to keep an informal look, or use shears for a more formal look.

**Zones:** 7 to 9

**Uses and Companions:** Combine Small Anise Tree with Azaleas, Rhododendrons, and deciduous shrubs. Planted in mass, it makes an effective evergreen screen. Use it where damp soils make it difficult to grow other shrubs.

## WAX MYRTLE
*Morella cerifera*

This native evergreen forms a large shrub or small tree that is easy to shape by pruning or shearing, depending on the desired effect. It provides both food and shelter for wildlife and makes a handsome screen or hedge. Trained as a small tree with a single trunk, it creates an interesting sculpture in the garden. Its aromatic foliage and colorful fruits add to this shrub's appeal. It also produces small grayish color fruits in fall. Once established, this shrub is drought friendly and will tolerate sandy soils and salt spray.

**Size:** It matures at 15 to 20 feet tall by 10 feet wide.

**Conditions:** Wax Myrtle likes full sun to part shade in an average soil, although it will tolerate sandy and moist soils.

**Zones:** 7 to 10

**Uses and Companions:** This tough, pest-free shrub is great for a screen or hedge. Use it with other native plants in natural settings. Combine it with native Azaleas and perennials like Cardinal Flower, *Lobelia cardinalis*.

# TEA OLIVE
*Osmanthus fragrans aurantiacus*

When Tea Olive is in bloom, its tiny flowers pump out sweet perfume that fills the air and delights the nose in spring, early summer, and fall. The glossy leaves, up to 4 inches long, are toothed or smooth. Perfect for a formal hedge, screen, or container plant, this large shrub or small tree adds elegance to the garden and requires no special care. There is also a form, *O. aurantiacus,* with fragrant orange flowers in fall.

**Size:** Tea Olive can grow to 15 to 30 feet tall by 15 to 20 feet wide.

**Conditions:** Full sun and well-drained soil are best for Tea Olive. Once established, this shrub is drought friendly.

**Zones:** 7 to 9

**Uses and Companions:** Use this evergreen for a backdrop, screen, or formal hedge. Plant deciduous spring-blooming shrubs like native Azaleas in front of it. Use it to divide different areas of the garden. Plant it near paths where you can appreciate its fragrance when it blooms. It also makes a good subject for containers.

# CHERRY LAUREL
*Prunus caroliniana*

This native evergreen tolerates wind, heat, and periods of drought. Fast growing, it makes an effective hedge, screen, or evergreen backdrop. As a large shrub, it can be branched from the base, but it can also be trained as a single-trunked tree. Small white flowers appear in early spring and are followed by black fruit, up to ½ inch wide. This shrub is easy to transplant and takes well to shearing.

**Size:** It grows 20 to 40 feet tall and 15 to 20 feet wide and has a pyramidal habit as a young tree, becoming more rounded as it matures. The cultivar 'Bright 'n Tight' is denser and more compact than the species, maturing at about 10 feet tall.

**Conditions:** Plant it in full sun, part shade, or shade in moist, well-drained soil.

**Zones:** 7 to 10

**Uses and Companions:** Use Cherry Laurel for formal or informal hedges and evergreen screens. 'Bright 'n Tight' also works well in containers. Combine Cherry Laurel with Rhododendrons, Azaleas, and deciduous flowering shrubs.

CANADIAN HEMLOCK

# MORE SHRUBS
## FOR SCREENING AND HEDGES

*Acer campestre*—Hedge Maple

*Aucuba japonica*—Japanese Aucuba

*Ilex glabra*—Inkberry Holly

*Loropetalum chinense*—Chinese Fringe

*Prunus laurocerasus*—English Laurel

*Rosa* 'Knock Out®'—Knock Out® Roses

*Thuja occidentalis* 'Emerald'—American Arborvitae

*Tsuga canadensis*—Canadian Hemlock

*Viburnum dilatatum* 'Erie'—Linden Viburnum

*Viburnum × pragense*—Prague Viburnum

# PUBLIC GARDENS AND ARBORETUMS

There are many wonderful public gardens across the United States. These living museums provide the opportunity to learn about plants that grow in a particular region, both native and exotic types. One definition for a public garden is "an area set aside for the cultivation of trees and shrubs for educational and scientific purposes." An arboretum focuses on woody plants, while botanical gardens include herbaceous plants as well as trees and shrubs. Some gardens have a special focus or collections that feature a group of plants, like native or rare and endangered species. Most public gardens offer educational programs and lectures to the public. The oldest public arboretum in North America is the Arnold Arboretum of Harvard University located in the Jamaica Plain section of Boston, Massachusetts. The arboretum occupies 265 acres and the living collections include over 7,082 accessioned plants. There are 4,544 botanical and horticultural taxa represented, with an emphasis on woody species of North America and eastern Asia.

KOREAN RHODODENDRON

Some of the major collections represented include Oak, *Quercus*; Beech, *Fagus*; Magnolia, *Magnolia*; Crabapple, *Malus*; Honeysuckle, *Lonicera*; Rhododendron, *Rhododendron*; and Lilac, *Syringa*. The Arboretum also maintains a herbarium with five million specimens and a library of more than 40,000 volumes.

LOROPETALUM CHINENSE
CHINESE FRINGE

# SHRUBS

Fall in the South can be a glorious time in the garden, and there are many shrubs that are bountiful, offering colorful foliage in shades of red, orange, yellow, and a blend of all three. Berries, too. Lavender, red, orange, white, and even black add a special touch. Knowing which shrubs put on an autumn show is part of creating a four-season garden. Combining these shrubs with trees, perennials, and bulbs culminates in a beautiful landscape with year-round interest.

With the arrival of cooler weather and (if we're lucky) some rain, fall is a good time to plant shrubs, trees, perennials, and bulbs in your landscape. Before you prune any fall-fruiting shrubs, find out if they produce fruit on new or old growth.

Evergreen shrubs, like Hollies, serve not only as hedges, backdrops, and specimens but also as a source for colorful berries in the fall and winter garden. Some provide food for wildlife too. Beautyberry, Viburnums, and Cotoneaster offer colorful fruits that brighten the fall garden. Many of these shrubs can be incorporated easily into the mixed border with perennials, ornamental grasses, and bulbs.

# AMERICAN BEAUTYBERRY
*Callicarpa americana*

Beginning in late summer or early fall, the native Beautyberry displays graceful, arching branches covered with spectacular lavender-colored fruits clustered around the leaf axils (where the leaves attach to the stem). Even after the leaves drop off, the bright-colored fruits persist. Another species called Purple Beautyberry, *Callicarpa dichotoma,* is also garden worthy. There are some selections with white fruit, but they don't add as much drama to the landscape as those with magenta or purple fruits. Beautyberry attracts birds, butterflies, and bees and, once established, will tolerate drought.

**Size:** *C. americana* can grow 6 feet tall. *C. dichotoma* grows 4 to 5 feet tall. In shade it will get taller.

**Conditions:** Plant Beautyberry in full sun or part shade in well-drained soil. Beautyberry blooms and fruits on current-year growth, so do any pruning in early spring before new growth starts.

**Zones:** *C. americana*: 7 to 10; *C. dichotoma*: 5 to 8

**Uses and Companions:** Beautyberry is ideal for the edge of the woodland or in the shrub border. Combine it with evergreen Hollies and the native Smoke Tree, *Cotinus obovatus*, for a beautiful autumn display.

# SUMMERSWEET
*Clethra alnifolia*

Summersweet offers fragrant spikes of flowers in summer and yellow to golden-brown foliage in fall. This native is easy to grow and adapts to a wide range of growing situations, including soils that are wet. A suckering shrub, it attracts bees, butterflies, and hummingbirds. You can use the cultivar called 'Hummingbird' as a groundcover under an evergreen shrub.

**Size:** A large shrub, Summersweet grows 6 feet tall by 5 feet wide. The cultivar 'Hummingbird' is a dwarf selection that has a compact and spreading habit, maturing at 3 feet by 3 feet.

**Conditions:** Plant Summersweet in full sun or part shade. It will grow in full shade, but the growth will be more open and leggy. This shrub will grow in average soils or wet soils.

**Zones:** 3 to 9

**Uses and Companions:** Plant Summersweet in a woodland garden at the edge of a pond. It also incorporates well into a collection of evergreen and deciduous shrubs.

# PARNEY COTONEASTER
*Cotoneaster lacteus*

This evergreen shrub has an open habit and attractive red fruits that persist into winter. An adaptable shrub, it is not bothered by extremes of heat, drought, or cold.

**Size:** This Cotoneaster has long, graceful, arching branches to 8 feet tall or more.

**Conditions:** Plant this shrub in full sun or part shade in well-drained soil.

**Zones:** 6 to 8

**Uses and Companions:** Train Parney Cotoneaster as an espalier, or use it on banks to help control erosion. Combine it with evergreen Hollies or conifers.

# MT. AIRY DWARF FOTHERGILLA
*Fothergilla gardenii 'Mt. Airy'*

It's hard not to notice this Fothergilla in autumn when its leaves turn brilliant shades of red, orange, and yellow. The honey-scented, bottlebrush-like flowers in spring are subtle but delightful when you get a whiff. In summer it displays handsome, blue-green witch hazel-type foliage.

**Size:** This vigorous form of Fothergilla can reach 6 to 8 feet tall at maturity.

**Conditions:** This shrub grows in part shade, but the fall color will be more intense if it is planted in full sun. Moist, well-drained soil is ideal.

**Zones:** 5 to 8

**Uses and Companions:** A plant for three seasons, 'Mt. Airy' is a suckering shrub, making it well suited for naturalistic plantings at the edge of a woodland, in a mixed border, or as a foundation plant. Planting it against an evergreen backdrop will highlight its blooms and foliage.

# CANARY AMERICAN HOLLY

*Ilex opaca* 'Canary'

Tough and serviceable, this native holly offers year-round interest in the garden with handsome evergreen foliage and colorful berries all winter, which also attract birds to your garden. It is important to note that most Hollies are dioecious, which means that male and female flowers are on different plants, but only the females bear fruit. To ensure good fruit production, you need to have a compatible male pollinator. It is best to purchase a male when you buy a female.

**Size:** Depending on the cultivar, American Hollies can reach anywhere from 3 to 40 feet tall at maturity. *Ilex opaca* 'Maryland Dwarf' is low and wide, perfect for bank plantings. Cultivars with red berries include 'Greenleaf', 'Jersey Princess', and 'Old Heavy Berry'. 'Canary' has yellow fruits.

**Conditions:** Plant this Holly in full sun or part shade in well-drained soil. Unlike some plants it is not bothered by air pollution.

**Zones:** 5 to 9

**Uses and Companions:** Evergreen Hollies make great hedges, screens, and specimen plants. Use them in combination with deciduous Hollies, native Azaleas, spring-flowering shrubs, and trees like Dogwoods and Redbuds.

# WINTERBERRY

*Ilex verticillata*

Winterberry, a deciduous Holly, is stunning in the winter landscape. Bright red fruits cover leafless branches and persist until birds eat them late in the season. They are also extremely hardy and will grow in a wide range of soil types. A

good male pollinator is 'Southern Gentleman'. 'Red Sprite' is a dwarf selection growing 3 to 4 feet tall; for a pollinator use 'Jim Dandy'. 'Sparkleberry' grows about 12 feet tall and is best pollinated by 'Apollo'. There are also cultivars with orange-red and yellow fruit.

**Size:** Depending on the selection, Winterberry can grow 6 to 10 feet tall.

**Conditions:** Plant deciduous Hollies in full sun or part shade for best fruit production. They will grow in wet or dry soils, but moist, well-drained soil is ideal.

**Zones:** 3 to 4

**Uses and Companions:** Use Winterberry for a focal point, in a group or in combination with evergreen Hollies and conifers like Blue Atlas Cedar.

# YAUPON HOLLY

*Ilex vomitoria*

Native to the South, Yaupon Holly is a versatile evergreen shrub with small leaves and tiny red berries. Whether you shear it, or use it as a screen or a focal point, it is easy to grow, attracts birds to your garden, and requires minimal care. Coastal gardeners will be glad to know that it tolerates salt spray.

**Size:** The species can grow 15 to 20 feet tall. There are also dwarf forms like 'Nana', which only grows 1½ feet high, and weeping forms like 'Pendula', which has long, weeping branches.

**Conditions:** This drought-tolerant species will grow in full sun or part shade. Yaupon grows in almost any soil.

**Zones:** 7 to 9

**Uses and Companions:** Use dwarf selections for hedges and foundation plantings. Weeping forms make great accent plants, and narrow, upright selections like 'Will Flemming' add height without taking up much width.

# HEAVENLY BAMBOO

*Nandina domestica*

This shrub has been a mainstay of Southern gardens for years, and with good reason. The handsome bamboo-like foliage and clusters of red berries from late winter until spring are great reasons to grow it. It tolerates drought, grows in sun or shade, and provides food for birds. Clusters of berries make great holiday decorations.

**Size:** The species grows 6 to 8 feet high and forms large clumps. There are numerous selections that only reach 2 feet or less at maturity.

**Conditions:** Plant Heavenly Bamboo in sun or shade in well-drained soil.

**Zones:** 6 to 9

**Uses and Companions:** Plant Heavenly Bamboo with other evergreen shrubs like Azaleas and Hollies. Some of the dwarf selections have colorful red foliage in winter. For a contrast, try the selection 'Alba', which has creamy-yellow berries.

# IROQUOIS VIBURNUM

*Viburnum dilatatum* 'Iroquois'

There are many great Viburnums, some known for fragrant flowers and others, like 'Iroquois', for its colorful and persistent fruit (in this case, red). Clusters of creamy-white flowers appear in late spring, and the rich green foliage looks good all summer. But fall is the season when this shrub shines. The leaves turn shades of red and orange before they drop and the fruit turns red, hanging onto the branches well into winter if the birds don't eat them. For the best fruit production, plant several Viburnums of the same species.

**Size:** Iroquois Viburnum has an upright habit, growing 9 feet tall by 5 feet wide.

**Conditions:** Plant this shrub in full sun or part shade in well-drained soil.

**Zones:** 5 to 7

**Uses and Companions:** This shrub is great for a mixed border with other shrubs and trees. Combine it with other Viburnums, Hemlocks or other evergreens, Florida Leucothoe, or Japanese Anise.

# TEA VIBURNUM

*Viburnum setigerum*

This is one of my favorite "fruiters" for fall. A loose, open shrub, Tea Viburunum produces long, arching branches covered with stunning orange-red fruits. Fall is also a time when the blue-green leaves take on shades of red. The cultivar 'Aurantiacum' has orange fruits. For the best show, plant one of each type.

**Size:** Tea Viburnum grows 8 to 12 feet tall.

**Conditions:** Plant this shrub in full sun or part shade in well-drained soil.

**Zones:** 5 to 7

**Uses and Companions:** I like Tea Viburnum planted in combination with other Viburnums and Purple Beautyberry, *Callicarpa dichotoma*. It also looks good against a backdrop of evergreen shrubs like Hollies.

SMOKETREE

# MORE SHRUBS
## FOR COLORFUL FALL FOLIAGE AND FRUIT

*Aesculus parviflora*—Bottlebrush Buckeye

*Calycanthus floridus*—Sweetshrub

*Cotinus obovatus*—American Smoketree

*Cotoneaster salicifolius*—Willowleaf Cotoneaster

*Ilex cornuta* 'Burfordii'—Burford Holly

*Itea virginica* 'Henry's Garnet'—Virginia Sweetspire

*Physocarpus opulifolius* 'Diablo™'—Ninebark

*Punica granatum* 'State Fair'—Hardy Dwarf Pomegranate

*Viburnum dentatum*—Arrowwood Viburnum

*Viburnum nudum* 'Winterthur'—Winterthur Viburnum

# BOTANICAL GARDENS OF THE SOUTH

When you are in the mood for a sunny day trip, here is a list of public gardens in the South that are sure to please and inspire you. See page 218 for a full listing of public gardens to be found in the South.

### Atlanta Botanical Garden

www.atlantabotanicalgarden.org

1345 Piedmont Ave. NE, Atlanta, GA 30309

(404) 876–5859

Call for hours of operation and admission fees. The garden contains thirty acres of gardens, a conservatory, an orchid collection, and more.

### Birmingham Botanical Gardens

www.bbgardens.org

2612 Lane Park Rd., Birmingham, AL 35223

(205) 414–3900

Open daily with free admission.

The garden has 67½ acres with 10,000 different plants and twenty-five unique gardens.

### Cheekwood Botanical Garden and Museum of Art

www.cheekwood.org

1200 Forrest Park Dr., Nashville, TN 37205

(615) 356–8000

Call for hours and admission fees.

There are fifty-five acres of gardens, including a Japanese garden; the museum displays American painting and sculpture.

### Crosby Arboretum

www.crosbyarboretum.msstate.edu

370 Ridge Rd. (mail: P.O. Box 1639)

Picayune, MS 39466

(601) 799–2311

Check for hours of operation and admission fees.

The arboretum features 104 acres dedicated to native plants of the Pearl River Drainage ecosystem.

### Daniel Stowe Botanical Garden

www.dsbg.org

6500 South New Hope Rd, Belmont, NC 28012

(704) 825–4490

Check for hours of operation and admission fees.

The garden contains 450 acres undergoing development as a botanical garden, including an orchid conservatory and many different plant collections.

### Riverbanks Zoo and Botanical Garden

www.riverbanks.org

500 Wildlife Parkway, Columbia, SC 29210

(803) 779–8717

The garden includes seventy acres of gardens with a variety of theme gardens.

### South Carolina Botanical Garden

www.clemson.edu/public/scbg/

Clemson University, 150 Discovery Lane, Clemson, SC 29634

(864) 656–3405

Open 365 days a year; there is an admission fee and a charge for some events.

There are 295 acres with 70 acres of demonstration and display gardens, a 40-acre arboretum, and 90 acres of woodlands, streams, managed meadows, turf, and shrubs.

*PHYSOCARPUS OPULIFOLIUS*
'DIABLO™' NINEBARK

# TREES

Selecting the right tree for the right place is especially important when you choose a shade tree for your garden. Unlike some shrubs and perennials, once you plant a tree, you don't want to move it. The more you know about its growth habit, anticipated mature size, and cultural requirements, the better prepared you will be to select the perfect trees for your landscape.

On my list of recommended shade trees I have included one ornamental flowering tree, *Prunus × yedoensis*, Yoshino Cherry. While it doesn't get as large as many shade trees, it will provide shade and is ideal for smaller gardens that need some shade but don't have room for larger trees.

When you plant large trees, keep in mind that if you can't find the selection you want in a large balled-and-burlapped size, container-grown trees are easier to transplant and will catch up in a few years with larger transplants.

While certain trees exhibit more drought tolerance than others, this is only after they become established in the landscape (usually after one or two years or growing seasons). Refer to the Introduction for more information about planting and care for trees.

# RED MAPLE
*Acer rubrum*

Native to low wet areas, Red Maple is a fast-growing tree that is attractive throughout the year. In late winter to early spring, red twigs and red flower clusters appear long before the leaves. New leaves emerge with tinges of red before they turn a rich green. The show continues in autumn with colorful foliage that ranges from yellow to brilliant red. Look for the selections 'October Glory' or 'Red Sunset', both noted for their brilliant orange and red leaves in fall.

**Size:** Give this tree plenty of space. It grows 40 to 60 feet tall and with a spread equal to or less than the height.

**Conditions:** Red Maple will tolerate a range of soil types but prefers slightly acidic soils and full sun or part shade.

**Zones:** 3 to 9

**Uses and Companions:** Red Maple has lots of surface roots; keep it away from paths and driveways where it could cause problems. It's a choice tree for a specimen, edge of the woodland, or meadow. Plant it against a backdrop of evergreens. Combine it with native Azaleas and other flowering shrubs.

# SHAGBARK HICKORY
*Carya ovata*

Shagbark Hickory has outstanding ornamental bark. Its "shaggy" look is due to the way thin strips of bark attach at the center to the trunk and curl away from the tree. The yellow-green leaves turn golden yellow in the fall when this Hickory also produces hard-shelled nuts that are filled with sweet fruit. This favorite native is also resistant to pest and disease problems.

**Size:** Shagbark Hickory grows 60 to 80 feet or taller. The leaves are 8 to 10 inches long, and the leaflets are ½ to 2½ inches wide.

**Conditions:** Full sun or part shade and well-drained soil are ideal. Shagbark will also tolerate drier soils. Because this native develops a large taproot, start with small container-grown plants.

**Zones:** 4 to 8

**Uses and Companions:** This beauty makes a statement as a specimen, in a group, or in combination with other trees. Plant it against a backdrop of evergreens to highlight the bark. Combine it with other native trees, shrubs, and ornamental grasses.

# WHITE ASH
### *Fraxinus americana*

White Ash is a fast-growing tree with a straight trunk and oval-shaped crown. It offers handsome dark green leaves, 8 to 15 inches long, with five to nine oval leaflets. In autumn the foliage turns shades of purple. It's a great shade or street tree that gives interest in the spring, summer, and fall. Look for seedless selections with long-lasting, purple fall foliage, like 'Autumn Applause' and 'Autumn Purple'.

**Size:** White Ash grows 50 to 80 feet tall with a similar spread.

**Conditions:** Full sun or part shade is best for White Ash, and it will tolerate average garden soils. To avoid maintenance headaches associated with heavy seed production, plant seedless varieties. Check to make sure your trees are healthy and free from borers or anthracnose before planting.

**Zones:** 3 to 9

**Uses and Companions:** Use White Ash as a specimen tree in the lawn, or plant it as a street tree. Plant a group of Ash trees to create shade. Combine it with evergreens such as American Holly.

# SWEETGUM
### *Liquidambar styraciflua*

Sweetgum is a tough tree that thrives with a minimum of care. Its handsome maplelike leaves, with five to seven lobes, start out green but turn purple, yellow, or red in the fall. The fruits, spiky balls called gumballs, hang on the tree during winter like ornaments. Gardeners who find the "Sweetgum balls" a nuisance may want to try 'Rotundiloba', a selection with rounded lobe leaves that does not set fruit. In its youth Sweetgum has an upright, cone-shaped habit. But as it ages, it spreads.

**Size:** Sweetgum matures to 60 to 75 feet with a spread of 20 to 25 feet.

**Conditions:** Full sun with moist, well-drained, slightly acidic soil are ideal, but Sweetgum will grow in a wide range of sites. Any pruning should be done in winter.

**Zones:** 5 to 9

**Uses and Companions:** Give Sweetgum plenty of room to establish roots. Use it as a lawn or street tree. Plant it near other shrubs and trees with outstanding fall color, such as Virginia sweetspire, *Itea virginica* 'Henry's Garnet'.

# TULIP POPLAR

*Liriodendron tulipifera*

This large native has handsome foliage, beautiful orange-and-green tulip-shaped flowers, and distinctive fruits that persist through winter. The leaves emerge bright green in spring, and in fall they turn yellow to golden yellow. Fast growing, Tulip Poplar forms a tall pyramidal crown. Selections like 'Arnold', with tapering, upright branches, work well where space is limited; it also blooms 2 to 3 years after planting, instead of the 10 to 12 years that it may take the species to bloom. Because of their shallow roots, it is best not to garden directly under Tulip Poplars.

**Size:** This tree grows 70 to 90 feet in height with a spread of 35 to 50 feet, but it can grow much taller.

**Conditions:** Full sun or part shade and moist, well-drained soil are best. Shelter this tree from wind, as it can be weak-wooded, with branches breaking up in storms or from ice.

**Zones:** 4 to 9

**Uses and Companions:** Plant majestic Tulip Poplars as specimens in the lawn or in groups where space permits. It's a good choice to create shade for smaller understory trees like Flowering Dogwood or Redbuds.

# SYCAMORE

*Platanus occidentalis*

With its ornamental bark and twisted branches, Sycamore looks like a living sculpture. The bark, which starts out a light grayish color, peels with age to expose patches of cream and white. The overall mottled effect, especially against blue skies, creates a sense of drama. This native makes a big impact, especially in winter and early spring before new leaves emerge. When the maplelike leaves, with three to five lobes, finally appear, they are up to 10 inches across. For large properties near streams, floodplains, or rivers, Sycamore offers year-round beauty.

**Size:** It grows 75 to 100 feet with an equal or greater spread.

**Conditions:** Sycamore likes full sun or part shade and rich, organic, moist, well-drained soil. Destroy dead leaves at the end of the growing season to reduce the potential for fungus spores (which cause dieback) to overwinter.

**Zones:** 4 to 9

**Uses and Companions:** Sycamore makes a grand specimen near a stream or river. A group of Sycamores creates a lasting impression. Underplant it with the shrub Virginia sweetspire, *Itea virginica* and Cardinal Flower, *Lobelia cardinalis*, a native perennial with bright red flowers in August.

# YOSHINO FLOWERING CHERRY

*Prunus × yedoensis*

Yoshino Flowering Cherry is a treat to experience in its spring glory. This tree is laden with fragrant, light pink—almost white—single blossoms. This fast-growing, graceful Cherry has handsome foliage, an upright, rounded habit, and horizontal branching. Not only is it beautiful, it tolerates our heat and humidity. In bloom, the cut branches make a wonderful addition to flower arrangements. The autumn foliage ranges in color from yellow to golden yellow.

**Size:** Yoshino Flowering Cherry reaches 30 to 40 feet tall with a 30 to 50 feet spread. The selection 'Akebono' reaches 20 to 25 feet tall and 25 feet wide.

**Conditions:** This tree likes full sun and moist, well-drained soil. Yoshino Flowering Cherry will tolerate clay soils. Prune only to remove errant branches.

**Zones:** 5 to 8

**Uses and Companions:** An allee or group of Yoshino Cherries in bloom is a glorious sight in early spring. Underplant this Flowering Cherry with early spring bulbs like small Daffodils and Grape Hyacinths or the Lenten Rose, *Helleborus orientalis.*

# WHITE OAK

*Quercus alba*

White Oak is a majestic giant that deserves a prominent spot in the landscape. The handsome foliage (4 to 8 inches long with rounded lobes), distinctive acorns, and winter silhouette all add up to a stately tree that adds beauty to every season. In its youth, this slow-growing native has a pyramidal habit, but as it matures it becomes broad and rounded with massive, wide-spreading branches. The foliage, which is dark green in summer, ranges from brown to wine red in autumn.

**Size:** White Oak matures to 50 to 90 feet tall with a spread of 50 to 80 feet, or wider.

**Conditions:** Full sun and a moist, well-drained, acidic soil are best. Prune in winter when trees are dormant. When planting an Oak, start with small trees. They are easier to establish.

**Zones:** 3 to 9

**Uses and Companions:** White Oak is ideal as a specimen tree for large landscapes. Plant this long-lived tree for future generations to enjoy. Avoid planting underneath Oaks. Plant this tree against a backdrop of evergreens like Hemlocks or American Hollies.

## LIVE OAK
*Quercus virginiana*

Live Oak is a choice evergreen for a shade tree. With its widespread, arching branches and a rounded canopy, this long-lived Oak makes a statement in the landscape. Not only is it beautiful, it adapts to a wide range of growing conditions, including moist and compacted soils. The new leaves (1¼ inches long and ⅜ to 1 inch wide) start out a bright olive color and mature to a lustrous dark green.

**Size:** Live Oak can reach 40 to 80 feet tall and 60 to 100 feet wide.

**Conditions:** Full sun or part shade and moist, well-drained soil are best for Live Oak, although it will grow in most soil types.

**Zones:** 7 to 10

**Uses and Companions:** Plant Live Oak as a specimen, street tree, or group of trees. Because this Oak tolerates salt spray, it makes a good shade tree at the beach. Combine it with native Azaleas and ornamental grasses.

## JAPANESE ZELKOVA
*Zelkova serrata*

A relative of the American Elm, Japanese Zelkova has a vase-shaped habit, handsome foliage, and, as it matures, ornamental, peeling, multicolored bark. Most important, this exotic tree grows happily in both urban and suburban settings without the problems that Elms usually suffer from and, once established, Japanese Zelkova will tolerate periods of drought. The dark green leaves of summer turn to shades of yellow, orange, and brown in autumn.

**Size:** Japanese Zelkova matures to 50 to 80 feet with an equal spread. 'Green Vase' and 'Halka' are selections for a vase-shaped habit.

**Conditions:** This tree likes full sun and moist, well-drained soil. Prune in fall to thin and shape. It's easy to transplant as a balled-and-burlapped tree.

**Zones:** 5 to 8

**Uses and Companions:** Japanese Zelkova makes an excellent street tree, specimen, or shade tree. Underplant with groundcovers like *Vinca minor*.

RED OAK

# MORE TREES
## FOR SHADE

*Acer saccharum*—Sugar Maple

*Carya glabra*—Pignut Hickory

*Fagus grandifolia*—American Beech

*Fraxinus pennsylvanica*—Green Ash

*Nyssa sylvatica*—Black Gum

*Quercus coccinea*—Scarlet Oak

*Quercus phellos*—Willow Oak

*Quercus rubra*—Red Oak

*Tilia americana*—American Linden

*Ulmus parvifolia*—Lacebark Elm

# RETURN OF THE AMERICAN ELM

At one time the American Elm, *Ulmus americana*, was a dominant feature in the American landscape, and Elm was also the most popular name for a street. Prior to World War II, the American Elm was the shade tree of choice throughout much of the United States. Large and graceful with an upright, vase-shaped habit, it was a popular street tree east of the Rockies. The drawback to planting so many of the same type of tree is that when they are affected by a blight or disease, as in the case with the Dutch Elm disease, great numbers of trees can be killed very quickly. In the 1930s, a fungal infection borne on beetles was introduced into the United States via a shipment of lumber from overseas. The results were devastating and tens of millions of American Elm trees were wiped out over a period of mere decades.

After twenty years of research, The National Arboretum has introduced at least two American Elms that are disease-tolerant (since no Elms seem completely resistant to Dutch Elm disease, or DED).

Both of these selections also exhibit a tolerance for air pollution and poor soils. If you love the American Elm, look for *Ulmus americana* 'Valley Forge' or 'New Harmony'.

For more information visit www.usna.usda.gov/Newintro/american.html.

QUERCUS PHELLOS
WILLOW OAK

# TREES

Fall is a special season to enjoy in the garden. It is a time to look forward to the colorful foliage that many different trees display before shedding their leaves for the winter. Subtle or bold, colors range from the richest reds and oranges to the palest yellows or brilliant combinations of colors. Some brighten the landscape for only a few days, while others entertain us for several weeks. One of the first trees to exhibit brilliant red and orange hues in its foliage is the native Tupelo. Its glossy leaves stand out in the landscape though they are here and gone all too soon. Maples, both native and Japanese types, are choice trees noted for their autumnal show.

Factors that influence the amount of fall color you will see include rainfall, drought, and temperature, and some types are definitely more ornamental than others.

Fall is a good time to plant many types of trees and shrubs in the South, and if you purchase them when they are putting on their autumn show, you will know you are getting a winner. The cooler weather makes it comfortable to tackle any of the tasks that await you in the garden.

Pairing the splendor of colorful trees with interesting shrubs and their fruits and foliage make for a fantastic display in the Southern, autumn garden.

# TRIDENT MAPLE

*Acer buergeranum*

This small- to medium-sized Maple is perfect where a small tree is needed. Undemanding and easy to transplant, it offers late-season color in autumn when other trees have already put on a show. Typically it has a rounded habit and can be grown as a large multi-stemmed shrub or a single-stemmed tree. The glossy foliage often starts out dark purple, changes to green in summer, and then changes to shades of red, yellow, orange, or all three in fall. As it matures the bark flakes off to reveal shades of gray-brown and orange.

**Size:** Trident Maples mature at 20 to 30 feet tall and almost as wide.

**Conditions:** Plant this tree in full sun or part shade in a well-drained soil. Once established, Trident Maple will tolerate periods of drought. It is not subject to any serious pest or disease problems.

**Zones:** 5 to 9

**Uses and Companions:** This small tree offers interest during spring, summer, and fall. Plant it near a patio as a street tree or specimen. Underplant it with groundcovers and ferns like Autumn Fern.

# JAPANESE MAPLE

*Acer palmatum* cultivars

Japanese Maples are elegant trees suited to formal gardens or to those that are more natural. Depending on the cultivar, the leaves vary from three-lobed to those that are finely dissected, almost like lace. They can have upright or weeping forms and some have colorful bark too. In autumn, certain selections turn brilliant shades of crimson, yellow, orange, and red. Coralbark Maple*, Acer palmatum* 'Sango-kaku', offers yellow foliage in autumn and striking coral bark in winter. 'Osakazuki' is noted for its vigor and outstanding scarlet foliage in autumn.

**Size:** This tree will grow 15 to 25 feet tall or taller and just as wide, depending on the selection.

**Conditions:** The ideal environment is moist, well-drained soil and full sun most of the day but offered protection from the hot, afternoon sun; filtered bright shade is good. Site so that they are not subject to sweeping winds or late frosts.

**Zones:** 5 to 8

**Uses and Companions:** Use Japanese Maples as specimens, accents, or as part of a mixed shrub border. They make good container plants. Plant against a backdrop of evergreens to help accentuate their striking forms in winter.

# RED MAPLE
*Acer rubrum*

Red Maple is a good choice for a fast-growing shade tree adapts to a range of soil types. In fall it is often one of the first trees to color, displaying scarlet leaves. There are numerous selections and hybrids available, noted both for their vigor and colorful autumn foliage. Noteworthy selections for the South include 'October Glory' with intense orange-red leaves in late fall and 'Autumn Flame®' with early red fall color.

**Size:** Depending on the cultivar, this tree grows from 40 to 60 feet tall with a spread that is less or equal to its height.

**Conditions:** Plant Red Maple in full sun or part shade. They will tolerate soils that are damp or subject to occasional flooding. Don't plant these trees too close to foundations, sidewalks, or driveways as they have many surface roots that will disrupt pavement.

**Zones:** 3 to 9

**Uses and Companions:** Use Red Maples as street trees, specimens in large areas, and as part of a mixed border with deciduous and evergreen trees. Combine them with native Azaleas and evergreen shrubs.

# PIGNUT HICKORY
*Carya glabra*

This native Hickory is a good choice for large properties where it can develop to its full potential. Pignut Hickory stands out in the autumn landscape when the leaves turn brilliant yellow, usually in late October to November. Tough and easy to grow, it is best to start with a young tree that is container grown. Hickories have a deep taproot and, once planted, should be left to grow and thrive. This species is not grown for its nuts, as the seeds are bitter.

**Size:** This tree has a broad canopy and grows 50 to 60 feet tall with a spread of 25 to 35 feet. The leaves, which contain five to nine leaflets, are 8 to 10 inches long.

**Conditions:** Ideal conditions include full sun, part shade, and well-drained soil. Once established, it will tolerate periods of drought.

**Zones:** 4 to 9

**Uses and Companions:** This native makes a grand specimen in large, open spaces. If space allows, combine it with other Hickories like the Shagbark Hickory.

# YELLOWWOOD
*Cladrastis kentukea*

Yellowwood is an elegant tree during every season. In spring the large foliage, 8 to 12 inches long and divided into seven to eleven oval leaflets, starts yellow-green before turning bright green in summer. Trees may not flower for ten years, but it is worth the wait. Clusters of creamy-white fragrant blooms, up to 14 inches long, resembling Wisteria, cover the tree from late spring to early summer. In autumn the leaves range from yellow to golden yellow. With the arrival of winter, we can appreciate its smooth, gray beech-like bark.

**Size:** Yellowwood grows 30 to 50 feet tall with a spread of 40 to 55 feet.

**Conditions:** Plant this deep-rooted tree in full sun or part shade in moist, well-drained soil. Prune to shape it as a young tree but do so only in summer, as it bleeds sap in winter and spring. Yellowwood tends to flower more heavily some years than others.

**Zones:** 4 to 8

**Uses and Companions:** Plant this tree as a specimen in the lawn, near the patio, or in a group with other Yellowwoods. Underplant it with ferns and shade-loving perennials like Hellebores.

# AMERICAN SMOKETREE 'GRACE'
*Cotinus obovatus* 'Grace'

The American Smoketree and hybrids like 'Grace' are rewarding and easy to grow as ornamentals. The "smoky" quality comes from the fluffy flower panicles in summer. The foliage on the American Smoketree starts out blue to dark green and turns brilliant red, yellow, and orange in fall. The foliage of 'Grace' starts out light red, turns dark red over summer, and then, bright orange in fall. Its flowers are a smoky pink. Both are effective as large shrubs or small trees.

**Size:** American Smoketree grows 20 to 30 feet tall for the species and 15 to 20 feet tall for 'Grace'.

**Conditions:** Ideal conditions include full sun or part shade and well-drained soil. Once established, Smoketree tolerates drought. Smoketree can be cut back hard every spring to keep it from getting too large (you'll sacrifice the flowers).

**Zones:** 5 to 9

**Uses and Companions:** Plant the American Smoketree as part of a mixed border with summer-blooming shrubs like *Hydrangea paniculata* 'Tardiva' and ornamental grasses. It also makes a good specimen. Plant it against a background of evergreens to highlight its colorful fall foliage.

# AMERICAN BEECH
*Fagus grandifolia*

This native makes a grand specimen with dark green summer foliage, golden-brown leaves in fall, and elephant-gray bark in winter. Many of the leaves persist through the winter (they look like thin silvery- brown tissue paper) until spring when they finally drop and new foliage emerges. The leaf buds and nuts are also ornamental.

**Size:** American Beech grows 50 to 75 feet tall and sometimes up to 100 feet tall with a spread that is equal to or less than the height of the tree.

**Conditions:** Full sun or part shade and moist, well-drained soil are best. This tree does not like compacted soils, and its shallow root system makes it difficult to grow grass or other plants around its base.

**Zones:** 3 to 9

**Uses and Companions:** American Beech makes a grand specimen for large properties. It is also an effective hedge or screen. Plant it in combination with evergreens like American Hollies or Hemlocks.

# GINKGO
*Ginkgo biloba*

This tree is beautiful and adaptable. The distinctive, fan-shaped leaves retain a bright green color all summer. But the real show is in autumn when the foliage turns golden yellow. When they shed their leaves, they create a gold carpet that lasts for weeks. Mature trees often develop a picturesque silhouette. This tree is dioecious, which means there are male and female trees. This is worth noting, because the females produce lots of foul-smelling naked seeds, making the males more desirable. Look for 'Autumn Gold', 'Princeton Sentry', and 'Fairmont'.

**Size:** Ginkgos usually grow 50 to 80 feet tall and 30 to 40 feet wide. Keep in mind, though, these trees can grow as tall as 100 feet.

**Conditions:** Plant it in full sun or part shade in well-drained soil. Ginkgo will tolerate air pollution, heat, and soils with a high salt content.

**Zones:** 3 to 8

**Uses and Companions:**
Ginkgo makes a good choice for urban landscapes in parks and as street trees, provided there is a big enough planting strip. Combine them with Red Maples and Tupelos for a spectacular fall display.

PROVEN PLANTS

173

# TUPELO OR BLACK GUM
*Nyssa sylvatica*

This native tree is one of the most consistent trees for outstanding fall color that ranges from scarlet to purple to orange to yellow. It is also one of the first trees to indicate that autumn has arrived when it displays one scarlet leaf. The glossy green summer foliage and striking winter silhouette are added reasons to grow this tree. As it matures the bark develops almost a scaly look.

**Size:** Tupelo grows 30 to 50 feet tall with a spread of 20 to 30 feet.

**Conditions:** Full sun or part shade and moist, well-drained soil are ideal. It has no serious pest or disease problems.

**Zones:** 3 to 9

**Uses and Companions:** This tree is a choice specimen or for a mixed planting in natural areas. Be sure to plant this tree in a location where you can appreciate its handsome foliage and winter habit. Combine it with other natives like Red Maple and Dogwoods.

This exotic is right at home in Southern gardens where, despite our heat, its foliage turns luminous shades of red, orange, and yellow in fall. Although awkward as a young tree, selective pruning can help it develop into a dense and shapely tree. The foot-long leaves are made up of ten to sixteen paired green leaflets, adding an interesting texture to the garden.

# CHINESE PISTACHE
*Pistacia chinensis*

**Size:** Chinese Pistache grows 30 to 60 feet tall with a similar spread.

**Conditions:** Full sun and well-drained soil are best, but this tree will tolerate a range of soil types, avoiding those that stay wet. Once established, it also tolerates periods of drought.

**Zones:** 6 to 9

**Uses and Companions:** It's a great tree for the home garden or city street, because Chinese Pistache tolerates heat, humidity, and poor soils. Combine it with other plants for fall like the Tea Viburnum, *Viburnum setigerum* or *Itea virginica* 'Henry's Garnet'.

SASSAFRAS

# MORE TREES
## FOR COLORFUL FALL FOLIAGE

*Asimina triloba*—Paw Paw

*Carpinus caroliniana*—Ironwood

*Carya ovata*—Shagbark Hickory

*Cercis canandensis*—Redbud

*Cornus florida*—Dogwood

*Diospyros virginiana*—American Persimmon

*Fraxinus americana*—American Ash

*Lagerstroemia indica*—Crapemyrtle

*Metasequoia glyptostroboides*—Dawn Redwood

*Oxydendrum arboreum*—Sourwood

# PLANTS WITH WET FEET

Sometimes the simplest and most effective solution for a problem area in our gardens is to work with the existing conditions rather than trying, usually with less-than-successful results, to change the environment. This is the best solution for areas where the soil stays saturated or is periodically flooded. There are a number of shrubs, trees, and herbaceous plants that are well equipped to grow in soil that is boggy one day and dry the next.

WEEPING WILLOW

## TREES

*Magnolia virginiana* — Sweetbay Magnolia

*Salix babylonica* — Weeping Willow

*Taxodium distichum* — Baldcypress

## SHRUBS

*Cephalanthus occidentalis* — Button Bush

*Clethra alnifolia* — Sweet Pepperbush

*Ilex glabra* — Inkberry Holly

*Ilex* × 'Sparkleberry' — Deciduous Holly

*Ilex verticillata* — Winterberry or Deciduous Holly (need a male and female plant for fruit production on the female)

*Itea virginica* 'Henry's Garnet' — Virginia Sweetspire

*Rhododendron viscosum* — Swamp Azalea

OXYDENDRUM ARBOREUM
SOURWOOD

Winter can be a lovely time in the Southern garden, especially with bare trees silhouetted against blue skies on a bright January or February day. Including trees in your garden with ornamental bark adds year-round interest, especially in winter. Some trees like Crapemyrtle and Kousa Dogwood offer not only handsome bark but beautiful blooms too. Winter King Hawthorn offers colorful bark and beautiful berries that persist into winter. An allée of 'Natchez' Crapemyrtles is a stunning sight no matter what the season.

When you plant trees with ornamental bark, think about siting them against a backdrop of evergreens or conifers, which will help to show off the bark.

Of those highlighted here, most become more ornamental as they mature. Some, like River Birch, which can get quite large, require lots of room to grow, have colorful peeling bark, and have a blend of colors that can be appreciated from the distance. Others, like the Paperbark Maple, have bark that is one color, in this case cinnamon, that peels off in thin layers. All of these trees make choice specimens or focal points in your Southern landscape.

# PAPERBARK MAPLE

*Acer griseum*

This handsome tree shines in the winter landscape when its cinnamon-colored bark peels off the trunk and branches in thin sheets. The green foliage (divided into three leaflets) turns brilliant red and orange in fall. It's truly a tree that provides four seasons of interest in the garden.

**Size:** Paperbark Maple matures at 25 to 30 feet or taller.

**Conditions:** Plant this Maple in full sun or part shade in moist, well-drained soil.

**Zones:** 5 to 8

**Uses and Companions:** Paperbark Maple provides the perfect focal point or specimen for small or large gardens. Plant it against a backdrop of evergreens to highlight the bark in winter.

# CORALBARK MAPLE

*Acer palmatum* 'Sango-kaku'

Coralbark Maple is a standout in the winter landscape with its coral-red bark, especially against blue skies. In spring and summer its delicate, bright green foliage give this vigorous tree a graceful feel. In autumn the leaves turn yellow, creating an interesting effect of red and yellow.

**Size:** Coralbark Maple grows 30 to 40 feet tall.

**Conditions:** Plant this Maple in full sun or part shade in moist, well-drained soil.

**Zones:** 5 to 8

**Uses and Companions:** Perfect as a specimen or a focal point, this beauty glows against a dark background of broadleaf evergreens like Hollies or conifers like *Cryptomeria* and *Chameacyparis obtusa*, Hinoki Cypress. Plant it in combination with ornamental grasses like *Carex* 'Indian Summer'. Underplant it with Autumn Fern, *Dryopteris erythrosora*, and Lenten Rose, *Helleborus* x *hybridus*, for an evergreen carpet.

# DURA-HEAT RIVER BIRCH

*Betula nigra 'Dura Heat™'*

*Betula nigra* 'Dura-Heat' offers fantastic bark year-round in shades of salmon, brown, cinnamon, and white. The bark is showy even on young trees. And, as the name implies, it is more heat tolerant and resistant to insect and disease problems associated with Birch trees. If you like the look of River Birch but don't have the space for a large tree, 'Little King' only reaches 10 to 12 feet high and 12 feet wide. With its rounded habit it makes a choice specimen or hedge.

**Size:** Dura-Heat River Birch grows 40 to 50 feet high and 30 to 40 feet wide.

**Conditions:** Plant this tree in full sun in moist soil.

**Zones:** 4 to 9

**Uses and Companions:** This is a great tree for areas where the soil stays wet for long periods. Combine it with other natives like Virginia Sweetspire, *Itea virginica* 'Henry's Garnet', Cinnamon Fern, *Osmunda cinnamomea*, and Cardinal Flower, *Lobelia cardinalis*. River Birch looks good in a group or as a specimen for large open areas.

As Kousa Dogwood matures, the bark becomes showy, exfoliating to reveal a mix of brown, gray, and tan. Besides the bark, stunning flowers, tapered white bracts that are 2 to 4 inches across, appear about three weeks after the native Dogwood, *Cornus florida*, blooms. The fruits are ornamental into fall unless birds or humans eat them. The leaves also turn shades of red and burgundy in the autumn, another plus for this multi-season tree. This species is also resistant to anthracnose, a disease that attacks the native Dogwood.

# KOUSA DOGWOOD

*Cornus kousa*

**Size:** Kousa Dogwood grows 20 to 30 feet tall and 15 to 20 feet wide. This tree has a vase-shaped habit as a young tree and becomes broader as it matures.

**Conditions:** Plant this tree in full sun in moist, well-drained soil.

**Zones:** 5 to 8

**Uses and Companions:** The Kousa Dogwood makes a great specimen or focal point. A group of three trees makes a stunning display in late spring. Plant them in combination with shrubs like Rhododendrons and Azaleas.

# CORNELIAN CHERRY DOGWOOD
*Cornus mas*

Attractive bark, showy yellow flowers in very early spring, red fruits in summer, and handsome leathery green foliage until late fall all add up to one great tree. This Dogwood is also tough and will grow in various types of soil. It suffers from no serious pest or disease problems and, once it is established, will tolerate a fair bit of drought. I had one that thrived for years on the narrow planting strip between the sidewalk and the street. Look for 'Spring Glow', a good choice for Southern gardens.

**Size:** Cornelian Cherry Dogwood grows 15 to 25 feet tall and 12 to 18 feet wide.

**Conditions:** Plant this Dogwood in full sun or part shade in well-drained soil.

**Zones:** 5 to 8

**Uses and Companions:** To highlight the flowers, plant this tree against a backdrop of evergreens. Group it with other Dogwoods like *C. florida* and *C. kousa* to have blooms from early spring through summer. Underplant it with early Daffodils and Hellebores for a spring show.

# WINTER KING HAWTHORN
*Crataegus viridis* 'Winter King'

In winter this multi-season tree is hard to miss. Even from a distance the brilliant red berries and the colorful, flaking bark, which ranges from silvery gray to orange on the inside layers of the trunk, make for a stunning display. The fruits persist late into the season until birds or other animals eat them. In spring the 2-inch-wide white flowers put on a show. Perhaps best of all, once

it is established, this disease-resistant, adaptable tree tolerates heat, drought, and compact soils, making it ideal for urban or suburban gardens.

**Size:** Winter King Hawthorn grows 25 to 35 feet tall and 20 to 25 feet wide, making it a small- to medium-sized deciduous tree.

**Conditions:** Plant Winter King Hawthorn in full sun or part shade in well-drained soil. This tree is also pollution tolerant, making it a good candidate for a street tree.

**Zones:** 4 to 7

**Uses and Companions:** Plant Winter King Hawthorn as a street tree, specimen in the lawn, or as part of a mixed border with evergreens. Plant it against a backdrop of evergreen Hollies or conifers for year-round interest.

# NATCHEZ CRAPEMYRTLE

*Lagerstroemia 'Natchez'*

This vigorous hybrid Crapemyrtle has been referred to as the "Queen of Crapemyrtles" and with good reason. Fantastic colorful bark, even on young plants, of cinnamon, white, and brown, a long season of beautiful white blooms (from June to September), as well as resistance to powdery mildew are what makes this tree so appealing. A bonus is its fall color, which can be orange or red.

**Size:** Natchez Crapemyrtle grows 20 to 30 feet tall and up to 35 feet wide, making it a large multi-stemmed shrub or a single-trunk small tree.

**Conditions:** A location of full sun and good air circulation prevents powdery mildew. Well-drained soil is best.

**Zones:** 7 to 9

**Uses and Companions:** Natchez Crapemyrtle makes a great street tree, an allée, lawn specimen, or patio tree. It also makes a good companion for summer blooming shrubs like Butterfly Bush and Purple Smoketree. For a dramatic effect, plant them in a group of three or more.

# TALL STEWARTIA

*Stewartia monadelpha*

Tall Stewartia is an elegant tree with wonderful, smooth, cinnamon bark that shines in the winter landscape. In early summer this small tree produces 1- to 1½-inch white, camellia-like blooms with yellow anthers. In autumn the rich green foliage turns shades of red to burgundy. Even the pointy buds are attractive on this exotic ornamental.

**Size:** This tree grows 20 to 25 feet tall. It is a popular subject for bonsai too.

**Conditions:** Full sun to part shade and moist, well-drained soil are ideal. While this Stewartia tolerates hot weather, it is happiest when it is sheltered from the hottest sun of the day. Be sure to apply several inches of mulch around the roots.

**Zones:** 6 to 8

**Uses and Companions:** Tall Stewartia makes a fine specimen, and it is great for a courtyard or as part of a mixed border with perennials and shrubs.

# JAPANESE STEWARTIA
*Stewartia pseudocamellia*

It's hard to say during which season this tree is at its best. In winter its silhouette and colorful mottled bark of gray, red, and brown create a striking scene. In summer its white, camellia-like flowers against green foliage, even in the bud stage, are beautiful. Then in autumn the leaves turn shades of red, orange, and yellow.

**Size:** Japanese Stewartia grows 20 to 40 feet tall and wide.

**Conditions:** This Stewartia will tolerate full sun but it will not tolerate drought or wind. For best results plant it in part shade.

**Zones:** 5 to 7

**Uses and Companions:** Japanese Stewartia makes a choice specimen or focal point. Underplant it with Ferns and Hellebores. Plant it against a background of Camellias.

# LACEBARK ELM
*Ulmus parvifolia*

Lacebark Elm is a graceful tree that provides four seasons of interest. It is also fast growing, tough, and disease resistant. In winter it offers a striking silhouette and its mottled bark flakes to expose shades of gray, green, orange, tan, and red. The bark becomes more ornamental as the tree ages. In spring small, bright green leaves cover the tree providing welcome shade. Then in autumn the leaves may turn shades of yellow, red, or purple. Leaf litter is at a minimum, because the small leaves are easy to maintain.

**Size:** Lacebark Elm grows 40 to 50 feet tall and 40 to 50 feet wide.

**Conditions:** Full sun and moist, well-drained soil are ideal, but Lacebark Elm will tolerate poor soils.

**Zones:** 5 to 9

**Uses and Companions:** Although it is not a replacement for the classic American Elm, *Ulmus americana*, the Lacebark Elm is a graceful tree with a broad, vase-shaped habit, perfect for a specimen, street tree, or shade tree. Plant it against a backdrop of evergreens like *Cryptomeria* or Arborvitae to highlight the ornamental bark.

HERITAGE™ RIVER BIRCH

# MORE TREES
## WITH COLORFUL BARK

*Acer buergerianum* — Trident Maple

*Betula nigra* 'Heritage™' — Heritage™ River Birch

*Carpinus caroliniana* — Ironwood

*Cladrastis kentukea* — Yellowwood

*Fagus grandifolia* — American Beech

*Firmiana simplex* — Chinese Parasol Tree

*Lagerstroemia fauriei* — Japanese Crapemyrtle

*Parrotia persica* — Persian Ironwood

*Platanus occidentalis* — Sycamore

*Pseudocydonia sinensis* — Chinese Quince

# THE TAP ROOT MYTH

Roots are the "heart" of a tree. The health of a tree is directly related to the health of its root system. Roots will only flourish in soils that have adequate amounts of water and oxygen.  It is a myth that most trees have taproots with the exception of trees such as White, Bur, and Black Oaks and Pecan.

When trees are young, they may have taproots, but as other types of roots begin to grow, the taproot is choked out. Trees may be killed as a result of compacted or flooded soils. These soils have little air, which impacts oxygen absorption and respiration in a tree. Here are some specifics on root types.

## TYPES OF ROOTS

**Taproots or "Sinker Roots"**—This is a large, main root located near the base of a tree that helps stabilize it, especially when the tree is young. They also serve to store carbohydrates and water.

**Lateral or Secondary Roots**—These roots grow horizontally out from the trunk. They often extend 2 to 3 times beyond the dripline (the point where the longest branches reach). They transport water and nutrients and provide perennial support. Trees with fibrous root systems and no taproot have a dense system of lateral roots.

**Absorbing Roots (Fine Roots)**—These roots are thin and short-lived; they are responsible for water and nutrient absorption. These roots are covered with microscopic root hairs to help with the uptake of water and are located in the top 10 to 12 inches of soil.

PLATANUS OCCIDENTALIS
SYCAMORE

The broad definition of conifers is "cone bearing" although there are conifers like Junipers and Yews that produce berrylike fruits. Worldwide there are over five hundred species of conifers. As ornamentals they offer year-round interest in the garden. Many are evergreen, with needle or scalelike leaves, but there are also deciduous types like the adaptable Dawn Redwood and Bald Cypress.

Conifers are not the first plant people think of for Southern gardens but there are many that are well adapted to our climate and they serve myriad roles in the landscape. Depending on the cultivar, they can be large or small and vary greatly in form and texture. Their colorful foliage is also noteworthy; there are types with blue, yellow, green, and a whole range of variegated forms.

Conifers such as Hemlocks and Cryptomeria provide effective screens, hedges, and backdrops, while Blue Atlas Cedar makes a beautiful focal point in the landscape. Dwarf selections can be incorporated into the mixed border to add welcome color, especially in winter, when many other deciduous plants have shed their leaves or died back to the ground. Conifers are also candidates for container gardening, on their own or combined with annuals and vines.

# BLUE ATLAS CEDAR

*Cedrus atlantica* 'Glauca'

Blue Atlas Cedar stands out in the landscape with its striking form and silver-blue needles. Tall and angular as a young tree, it becomes less open as it matures. Like other cedars it stands up to heat and humidity and, once established, is drought tolerant. Pruning or pinching back young shoots will help keep branches from getting too heavy.

**Size:** As a young tree, Blue Atlas Cedar grows at a moderate pace and will grow to 60 feet or taller. Weeping forms like 'Glauca Pendula' and 'Pendula' have vertical drooping branches. The main trunk can be staked to a desired height or left to ramble, acting more like a groundcover.

**Conditions:** Plant Blue Atlas Cedar in full sun in moist, well-drained soil.

**Zones:** 6 to 9

**Uses and Companions:** Blue Atlas Cedar makes a spectacular specimen or focal point in the landscape. Combine it with ornamental grasses, perennials, and broadleaf evergreens like *Osmanthus fragrans* or Hollies.

# DEODAR CEDAR

*Cedrus deodara*

This conifer is an elegant evergreen that offers color, texture, and form year-round. Native to the Himalayas, it grows happily in our heat and humidity. As a young tree it has smooth bark and a pyramidal shape with stiff branches. The sharp, pointed 1- to 2-inch-long dark green needles sometimes have a silvery blue-green cast. As Deodar Cedar matures, its branch tips droop and it develops into a wider, more graceful tree with furrowed, scaly bark. The decorative, upright cones start out purplish green and turn reddish brown with age.

**Size:** While the species can grow to be 60 feet or taller in the garden, there are numerous selections that are slower growing and more compact like *Cedrus deodara* 'Divinely Blue' with blue needles.

**Conditions:** Plant this evergreen in full sun and well-drained soil.

**Zones:** 7 to 9a

**Uses and Companions:** Plant Deodar Cedar as a specimen, a screen, or background for deciduous trees and shrubs like Japanese Maples and Red Twig Dogwoods. Use dwarf selections in the perennial border or large pots to provide winter color.

# PROSTRATE JAPANESE PLUM YEW

*Cephalotaxus harringtonii 'Prostrata'*

This low-growing evergreen makes a choice foundation plant and thrives in shade or part sun. The Japanese Plum Yew looks like a Yew *(Taxus)* but is a much better choice for our humid Southern gardens. It is well mannered and elegant in the landscape. It makes an effective groundcover in open woodland settings in combination with deciduous shrubs like native Azaleas.

**Size:** The slow-growing cultivar 'Prostrata' grows 3 to 4 feet tall by 3 to 4 feet wide in about ten years. Other forms include 'Fastigiata', which is upright and narrow, growing to about 10 feet tall by 6 to 9 feet wide.

**Conditions:** A moist, well-drained soil is best when the plant is young. Once established, it can tolerate some drought. Plant in shade or part sun.

**Zones:** 5 to 9

**Uses and Companions:** Prostrate Japanese Plum Yew makes a great foundation plant where evergreen plants are needed that won't require constant pruning. For a contrast, combine it with Cast Iron Plant, which has broad, tall leaves.

# HINOKI FALSE CYPRESS

*Chameacyparis obtusa*

Flat, fanlike sprays of compressed emerald-green foliage make Hinoki False Cypress one of the most elegant evergreens, both in the landscape and in containers. Among the many cultivars I am drawn to are the dwarf selections 'Nana Gracilis' and 'Nana Lutea' with foliage that is green and gold. *Chameacyparis obtusa* 'Nana Lutea', known as Gold Dwarf False Cypress, is very slow growing and takes years to reach 4 to 6 feet tall by 4 feet wide.

**Size:** The species *Chameacyparis obtusa* is great if you have room, as it reaches 50 to 75 feet by 10 to 20 feet wide at maturity.

**Conditions:** Plant Hinoki False Cypress in full sun or part shade in moist, well-drained soil. Golden-leaved or variegated selections need full sun to maintain color; if they are planted in shade the leaves revert to green.

**Zones:** 4 to 8

**Uses and Companions:** While larger selections are great for screening, an evergreen backdrop, or a specimen, dwarf cultivars are perfect for containers providing low-maintenance, year-round color. Adding annuals like Pansies for winter and *Calibrachoa* for summer will dress up your pots.

# JAPANESE CRYPTOMERIA

*Cryptomeria japonica*

For an evergreen screen or specimen, Japanese Cryptomeria is hard to surpass. In summer it grows at a moderate speed with a pyramidal habit and smooth, compressed, rich green needles. In cold winters the foliage may turn bronze green. Once established, *Cryptomeria* can tolerate some drought. 'Black Dragon' has foliage that starts out green in spring and is very dark (almost black) the rest of the year.

**Size:** The species can reach 30 to 50 feet tall or taller and 12 to 20 feet wide at the base. *Cryptomeria japonica* 'Globosa Nana' grows 2 to 4 feet tall by 2½ to 3½ feet wide.

**Conditions:** Plant Japanese *Cryptomeria* in full sun or part shade in moist, well-drained soil. A location with good air circulation will cut down on disease. Prune dead or dying branches to prevent the spread of disease.

**Zones:** 5 to 9

**Uses and Companions:** Japanese *Cryptomeria* is ideal for screening, a hedge, or a specimen tree. Use it as a backdrop for deciduous trees like Dogwoods, Redbuds, or Japanese Maples.

# EASTERN RED CEDAR

*Juniperus virginiana*

This native plant is a familiar sight growing in fields and along highways throughout the Southeast. This tough guy is garden worthy for its handsome dark green foliage and reddish-brown bark. The scalelike needles range in color from green to steel blue and in winter take on shades of purple. Eastern Red Cedar grows in a wide range of soils and will tolerate drought and heat. The cones resemble berries and start out green in summer and turn dark blue in autumn.

**Size:** The species grows 40 to 50 feet tall with a conical shape. 'Grey Owl' is a compact form with silvery-blue-gray needles; it grows 2 to 3 feet high and 4 to 6 feet wide. Upright types include 'Canaertii' and 'Burkii'.

**Conditions:** Plant Eastern Red Cedar in full sun in well-drained soil. It is subject to cedar apple rust and should not be grown in the same area with Apples.

**Zones:** 2 to 9

**Uses and Companions:** Use Eastern Red Cedar as a windbreak, screen, or evergreen in natural areas where it will provide food and shelter for birds. Combine it with other natives like American Holly, *Ilex opaca*.

# DAWN REDWOOD
*Metasequoia glyptostroboides*

A stately tree with few demands, Dawn Redwood is native to China and Japan. This deciduous conifer thrives in many different garden settings. Fast growing and adaptable, its small, soft green needles give the tree a delicate look. In autumn, the needles turn yellow to russet red to bronze before they drop off to highlight its conical habit, horizontal bare branches, and distinctive bark. The tiny ½- to 1-inch-long cones are also decorative.

**Size:** A large tree, Dawn Redwood grows 75 to 100 feet tall and 20 to 30 feet wide at maturity. The selection 'Ogon' has golden foliage that turns orange in the autumn.

**Conditions:** Plant this conifer in full sun or part shade in well-drained soil. It requires no pruning or special care.

**Zones:** 4 to 8

**Uses and Companions:** Plant Dawn Redwood as a single specimen or in a group for a big impact. Combine it with evergreen conifers or Hollies. Plant it in a location with lots of room.

# BALD CYPRESS
*Taxodium distichum*

Bald Cypress is native to the Southeast. Although it is large, its flat, needlelike leaves (about ½ inch long) and 1-inch cones give it a graceful feel. A deciduous conifer, its light green leaves turn shades of orange, red, and brown before shedding in autumn to reveal a beautiful branching structure and peeling, cinnamon bark. Its ability to adapt to a wide range of growing sites is another reason for its popularity. I have seen Bald Cypress thriving in swamps and in parking lots surrounded by asphalt.

**Size:** As Bald Cypress ages it becomes broader at the top, maturing at 50 to 70 feet or taller. A weeping form, 'Cascade Falls', takes up much less space depending on how high it is staked, but offers the same delicate look.

**Conditions:** While Bald Cypress will grow in most soils, it does not like alkaline soils. It will grow in full sun or part shade.

**Zones:** 4 to 9

**Uses and Companions:** Bald Cypress will thrive at the edge of a pond, stream, or in a swamp. It also makes a good choice for a grove or specimen. Pair it with *Itea virginica* 'Henry's Garnet' or Water Iris.

# GREEN GIANT ARBORVITAE

*Thuja* 'Green Giant'

Green Giant Arborvitae is fast growing, adaptable, and useful as an evergreen screen, windbreak, or specimen in the landscape. The handsome, flat sprays of bright-green foliage make it a winner year-round. It is also resistant to pest and disease problems and naturally forms a pyramidal shape, which requires very little pruning.

**Size:** 'Green Giant' grows 50 to 60 feet tall and 20 feet wide, so give it lots of room. Space them 15 feet apart for the best screening.

**Conditions:** Plant this Arborvitae in full sun in well-drained soil.

**Zones:** 5 to 7

**Uses and Companions:** 'Green Giant' is perfect for a mass planting or as a single specimen. Use it for a backdrop for deciduous shrubs like Viburnums or Roses.

# CANADIAN HEMLOCK

*Tsuga canadensis*

This native hemlock responds well to heavy pruning and is equally suited for formal clipped hedges, informal screening, or as a specimen. The dark-green needles have a distinctive white band on the undersides that adds to their appeal. In a drought situation it is more susceptible to insect and disease problems.

**Size:** A dense pyramidal tree, hemlocks grow 40 to 70 feet or taller. There are dwarf and weeping forms available that take up much less space, including 'Sargentii' that can be staked to 15 feet. It will probably still grow 20 to 30 feet wide at maturity.

**Conditions:** Hemlocks like moist, well-drained soil and protection from hot afternoon sun. They are an ideal choice for a woodland garden.

**Zones:** 4 to 7

**Uses and Companions:** Hemlocks are happiest in woodsy situations with Mountain Laurels, Rhododendrons, and native Azaleas. They are also well suited for a clipped hedge in more formal settings.

LACEBARK PINE    MORE CONIFERS

*Calocedrus decurrens* — California Incense Cedar

*Chameacyparis pisifera* — Sawara False Cypress

*Cunninghamia lanceolata* — Chinese Fir Tree

*Cupressus arizonica* — Arizona Cypress

*Picea abies* — Norway Spruce

*Pinus bungeana* — Lacebark Pine

*Pinus taeda* — Loblolly Pine

*Pinus thunbergii* — Japanese Black Pine

*Thuja orientalis* — Oriental Arborvitae

*Thujopsis dolobrata* — Broad-Leaved Arborvitae

# HARDY PALMS

The good news is you don't have to live in Florida to grow Palms. If you want to add a tropical feel to your landscape, there are a number of hardy Palms that thrive in many parts of the South. Here are a few that will be happy in your garden.

NEEDLE PALM

*Rhapidophyllum hystrix*, **Needle Palm**—This adaptable native grows in sun or shade and gets its common name from the sharp needles that protect its crown. Extremely hardy (to Zone 6b) it is right at home in the Southeast with our warm, moist summers. A clumper with many large palmate leaves, it grows slowly and can reach 10 feet high and wide, although but 5 feet high and wide is more typical.

*Sabal minor*, **Dwarf Palmetto**—This Palm tolerates a good bit of shade or sun. It grows 4 to 5 feet high and wide. It likes a moist soil and sun, making it a great choice to site near a lake or on the bank of a creek.

*Trachycarpus fortunei*, **Windmill Palm**—This Palm originating from eastern China grows with a single trunk and about twenty fan-shaped leaves. In the southeastrn United States, it grows about 25 feet high. Plant it in full sun or light shade.

For more information, contact the Southeastern Palm Society at www.sepalms.org.

The Southeastern Palm Society is the south-eastern United States (north-of-Florida) chapter of the International Palm Society. SPS members are devoted and enthusiastic growers of hardy palms and other subtropical plants. Members enjoy sharing experiences through the society's quarterly journal, *Southeastern Palms*, and at quarterly meetings.

*Southeastern Palms* typically has twenty pages and features extensive use of color, with profiles of hardy palms and other subtropical plants and many backyard reports that details the experiences and growing tips of SPS members. The Society occasionally publishes other unique gardening guides, such as *Hardy Citrus for the Southeast* and *Hardy Palms for the Southeast*.

CUPRESSUS ARIZONICA
ARIZONA CYPRESS

# FLOWERING BULBS
## FOR SPRING

For many gardeners, Daffodils or Jonquils, common names for members of the genus *Narcissus*, herald the arrival of spring in the South, but there are many other bulbs to brighten the spring scene. Hardy spring-blooming bulbs range in size from the small species *Crocus*, just a few inches high, to large Hardy Amaryllis, up to 2 feet tall, with baseball-sized red flowers.

The best time to plant these spring bulbs is in mid-fall, usually October or November, once soil temperatures have cooled. Well-drained soil that has been amended with organic matter is ideal. Most of these require full sun, but there are selections like Summer Snowflake, *Leucojum aestivum*, which grow in sun or shade, and they tolerate damp soils too.

Many of these bulbs will perennialize and bloom year after year. You can plant one type in large groups, or you can combine different varieties and species to create a dazzling display. Smaller types can be incorporated into the perennial border, and many are good candidates for naturalizing. If you garden in containers, you can layer bulbs in a large pot for a succession of blooms. With careful planning you can have blooms from January through spring.

# GLORY OF THE SNOW

*Chionodoxa luciliae (gigantea)*

This charming heirloom dates to 1878. Starlike lavender-blue blooms with bluish-white centers grace the garden from late winter to early spring. There is also a white form called 'Alba'. Glory of the Snow is great for naturalizing in lawns, woodlands, or rock gardens. You can force this bulb in pots on its own or with other bulbs.

**Size:** Glory of the Snow ranges from 5 to 10 inches tall.

**Conditions:** Plant Glory of the Snow 5 inches deep from the base of the bulb and 1 inch apart in full sun or part shade. Although it will tolerate periods of drought once established, starting with moist, well-drained soil will give your bulbs a jumpstart.

**Zones:** 3 to 8

**Uses and Companions:** Plant drifts of bulbs in the woodland or lawn. Combine them with larger bulbs like Daffodils or Tulips. Plant them under Flowering Quince, native Azaleas, or Witch Hazels to create a carpet of bloom. Hellebores make good companions too.

# CROCUS

*Crocus* spp. and cultivars

For color in late winter or early spring, Crocus are hard to beat. With their beautiful cup-shaped flowers and grasslike foliage, these small bulbs create a carpet of bloom. They are great for naturalizing or forcing in pots. *Crocus chrysanthus* 'Snow Bunting' has pure white flowers with dark lilac feathering and bronze in the center. There are also selections with cream, yellow, blue, and purple flowers. Squirrels and chipmunks consider these bulbs a gourmet treat, but the good news is that *Crocus tommasinianus,* although smaller, is vigorous and resistant to squirrels.

**Size:** Depending on the selection, Crocus grow 3 to 6 inches high.

**Conditions:** Plant Crocus bulbs in fall. Mark the areas where you plant them so you can fertilize next fall after the foliage has died down. If you plant Crocus in the lawn, wait six weeks after they bloom before mowing. Even then, set your mower on the highest setting.

**Zones:** 3 to 8

**Uses and Companions:** Tuck groups of Crocus into the area around trees, in the lawn, or in pots with other spring-flowering bulbs like Daffodils and Tulips. Because they are small, you can tuck them in among tree roots.

# AMARYLLIS

*Hippeastrum × johnsonii*

Hardy Amaryllis is an old-fashioned favorite that lights up the spring garden with its bright red trumpets marked with white on the inside. There can be as many as four stems per bulb and six flowers per stem. The wide, green, straplike foliage looks good all summer and can take on shades of copper in the autumn.

**Size:** Hardy Amaryllis grow 20 inches tall and clumps can be up to 2 feet across.

**Conditions:** Plant Amaryllis in spring or fall. Good drainage in winter is important to ensure hardiness. Plant them in full sun or part shade. The tips of the bulbs should be just slightly covered with soil.

**Zones:** 7 to 11

**Uses and Companions:** Plant this bulb in the perennial border with other red flowers and foliage like Smoke Tree, Red Verbena, and Japanese Maple.

# BLUEBELLS OR WOOD HYACINTH

*Hyacinthoides hispanica* 'Excelsior'

Not only does this heirloom bloom in sun or shade, it also naturalizes freely. With rich, deep blue, bell-shaped flowers and glossy foliage, this bulb is a winner in the perennial border or woodland. It also makes a choice cut flower with 8- to 12-inch stems. Selections include 'Queen of the Pinks' with soft lavender-pink flowers and 'White City' with snow-white flowers. Another common name for Bluebells is Wood Hyacinths.

**Size:** Bluebells grow 8 to 12 inches tall depending on the selection.

**Conditions:** Plant bulbs in moist, well-drained soil in sun or shade.

**Zones:** 4 to 10

**Uses and Companions:** Combine Bluebells with Ferns, Hellebores, and *Ipheion uniflorum*. Plant them under Azaleas, Rhododendrons, and the Chinese Snowball Viburnum, *Viburnum macrocephalum*. They also look good planted in the woodland.

# HYACINTH
*Hyacinthus orientalis*

Fragrance, fragrance, fragrance! This flower is not graceful, but it is colorful and sweet, a true sign of spring. The blooms can be single or double, with petals that curve back. Hyacinths come in a range of colors including white, blue, purple, pink, and all the variations of these colors. Because they are easy to force, you can create a fragrant pot garden. They are also pest resistant, a plus for gardeners plagued by squirrels and other critters.

**Size:** Hyacinths display 8- to 10-inch stems covered with florets that add up to large flower spikes.

**Conditions:** Plant in fall with other spring-flowering bulbs but not too early as the bulbs need to be completely dry before you plant them. Plant them in full sun in moist, well-drained soil spaced at five per square foot.

**Zones:** 4 to 8

**Uses and Companions:** Their stiff, upright form make Hyacinths well suited for formal gardens and for patterned plantings. They look good in combination with Daffodils, Tulips, *Muscari*, Anemone, and *Chionodoxa*, bulbs that bloom at the same time. Other companions include Euphorbias and Violas.

Star Flower has been grown in

# STAR FLOWER
*Ipheion uniflorum*

Southern gardens for years. An heirloom that dates to 1832, it's no wonder that it is still popular today. Fragrant star-shaped flowers that range from almost white to violet, a long bloom period, and resistance to pests all contribute to its success. Star Flower naturalizes easily in lawns but is happy in the flower border too. Plant bulbs in large groups of twenty or more to make an impact. The flowers smell sweet, but if you crush the foliage, it gives off a distinct garlic odor.

**Size:** Although it is only 2 to 3 inches tall, this bulb spreads quickly to create a mass of blooms sometimes covering an entire lawn.

**Conditions:** Plant Star Flower in full sun or part shade in well-drained soil. If they are planted in the lawn, wait five or six weeks until after they bloom before you mow and cut back the foliage.

**Zones:** 5 to 9

**Uses and Companions:** Use Star Flower for early spring bloom in the lawn, woodland, or flower border. Combine it with Hellebores and early Daffodils. Use it as a groundcover under native Azaleas and shrubs like *Kerria japonica* with its bright yellow flowers.

# SNOWFLAKE OR SUMMER SNOWFLAKE

*Leucojum aestivum*

This versatile bulb is a must for every garden in the South. Whether you have a boggy site or moist, well-drained soil, Snowflake grows in sun or half shade. White, bell-shaped pendant flowers with green tips sway in the breeze and bloom for weeks beginning in mid-spring. The foliage looks good for months. Critters don't bother this bulb, and maybe that is why it naturalizes so readily. 'Gravetye Giant' is known to have larger flowers and be more robust than the species, but both are winners.

**Size:** Snowflake matures at 12 to 18 inches tall; 'Gravetye Giant' grows 18 to 24 inches tall.

**Conditions:** Plant Snowflake in fall in full sun or part shade. It will tolerate a wide range of soil types including damp soils. After it blooms, let the foliage ripen until one-third of it turns yellow.

**Zones:** 4 to 8

**Uses and Companions:** Use Snowflake with other bulbs like Daffodils, Muscari, and Tulips. Combine it with Hellebores, Christmas Ferns, and Hardy Gingers. Plant it next to Hydrangeas and native Azaleas. A bog or the edge of a pond is another good spot for these tough guys.

Sometimes small bulbs put on a big show. This is the case with Grape Hyacinth. In early to late spring, cobalt-blue spikes of grapelike, scented blooms grow only 4 to 6 inches tall, but wow! Resistant to four-legged pests, Grape Hyacinth persists in the garden for years. Unlike most spring bloomers, Grape Hyacinth foliage appears in autumn so you can use it as a marker to indicate where other spring bloomers that benefit from fertilizing in fall are planted.

# GRAPE HYACINTH

*Muscari armeniacum*

**Size:** *Muscari* has tiny spikes of tight blooms that grow 4 to 6 inches tall.

**Conditions:** Plant Grape Hyacinth in fall in moist, well-drained soil in full sun or part shade. Plant them in large groups of twenty or more bulbs.

**Zones:** 4 to 9

**Uses and Companions:** Plant them in the front of the border, in the lawn, or woodland. They are a good choice for naturalizing, forcing, or tucking under trees and shrubs or in the border. Combine them in pots with Tulips and Daffodils. Virginia Bluebells, *Mertensia virginica*, is another good companion for this charming bulb.

# DAFFODIL
*Narcissus* spp. and cultivars

Daffodils (and its relative, Jonquils) with their cheerful flowers of yellow, white, orange, pink, and all combinations thereof, are a sure sign that "spring has sprung." Many selections are fragrant and easy to force. There are dozens of types, from the small *Narcissus* 'Tete-a-Tete' to the large 'Carlton' growing 18 to 20 inches high. Once established, Daffodils persist for years and require minimal maintenance. Because they are poisonous, critters don't seem to bother them. With some planning it is possible to have Daffodils blooming from February until May.

**Size:** Daffodils range from 5 to 6 inches tall to types growing 15 to 17 inches tall.

**Conditions:** Most Daffodils require full sun and well-drained soil. Let at least one-third of the foliage turn yellow before you cut it back. Fertilize Daffodils in fall with a slow-release fertilizer with a formula of 5-10-20 (Daffodils like extra potassium).

**Zones:** 3 to 8. Tolerance may vary by cultivar. Some are hardier than others.

**Uses and Companions:** Combine Hellebores and Hostas with Daffodils, as they will help mask the ripening foliage. You can layer bulbs so that Daffodils, Tulips, and Hyacinths bloom at the same time to create a fantastic effect. Choose varieties that naturalize easily for mass plantings.

# TULIP
*Tulipa* spp. and cultivars

Tulips bring excitement to the garden in a wide range of often intense colors. But there are smaller, more delicate looking species that, while not as showy, persist in the garden. With fifteen different divisions based on shape, bloom time, and heritage, it's hard to choose a favorite. 'Apricot Beauty', a fragrant, single, early variety, is a soft salmon color with light rose flames on the outer petals; it is great for forcing or the border. *Tulipa clusiana* 'Cynthia' has red petals edged with chartreuse and a purple base and is a strong grower.

**Size:** Depending on the selection, Tulips can grow 4 to 6 inches tall or 20 to 24 inches tall. The flowers can be tiny or big and bold.

**Conditions:** Plant Tulips in fall with other spring-blooming bulbs. Plant medium- and larger-sized Tulips at a depth four times their height, 8 to 10 inches. Plant smaller types 4 to 6 inches deep. Both prefer full sun in moist, well-drained soil.

**Zones:** 3 to 8–9. There is some variance with different selections.

**Uses and Companions:** Combine Tulips with Daffodils, Scillas, *Ipheion*, and *Muscari*. Dogwoods and Azaleas also make happy companions for many types of Tulips. For formal bedding the Darwin Hybrid Tulips are considered a long-lasting perennial type.

DWARF IRIS

# MORE BULBS
## FOR SPRING

*Allium giganteum*—*Ornamental Onion*

*Anemone coronaria*—*Windflower*

*Camassia leichtlinii* 'Caerulea'—*Camassia*

*Eremurus* Spring Valley Hybrids—*Foxtail Lily*

*Galanthus nivalis*—*Snowdrop*

*Gladiolus byzantinus*—*Byzantine Gladiolus*

*Fritillaria meleagris*—*Guinea Hen Flower*

*Iris reticulata*—*Dwarf Iris*

*Puschkinia scilloides* var. *libanotica*—*Puschkinia*

*Scilla siberica*—*Scilla*

# HEIRLOOM BULBS

According to Scott Kunst, founder and owner of Old House Gardens, "it's not just in the rainforest" that unique, valuable plants are being lost forever. He believes that antique bulbs are as much a part of our cultural identity as Colonial houses, pioneer quilts, Victorian rockers, and 1950s Chevrolets. In the South there are many heirloom plants and bulbs that once graced our gardens but have disappeared and are hard to find in the trade. An heirloom plant is defined in a number of different ways, but generally it is considered to be a cultivated variety of a flowering plant that has been grown for at least fifty years. Fortunately, there are a number of mail-order sources for some of these treasures. There are many heirloom varieties of Daffodil, *Narcissus*, that are easy to grow and, perhaps because the bulbs are poisonous, not appealing to critters.

Below is a list of Daffodils that Old House Gardens recommends for Southern gardens. All of the following heirlooms are fragrant except for 'Dick Wellbrand' and 'St. Keverne'.

| | | |
|---|---|---|
| 'Avalanche' | 'Geranium' | 'Sweetness' |
| 'Dick Wellbrand' | 'Grand Primo' | 'Texas Star' |
| 'Early Pearl' | 'Early Louisiana' | 'Thalia' |
| 'Erlicheer' | 'St. Keverne' | 'Trevithian' |

## SOURCES FOR HEIRLOOM BULBS

Brent and Becky's Bulbs
7900 Daffodil Lane
Gloucester, VA 23061
877-661-2852
www.brentandbeckysbulbs.com

Old House Gardens
536 Third St.
Ann Arbor, MI 48103
734-995-1486
www.oldhousegardens.com

The Southern Bulb Co.
www.southernbulbs.com

*TULIPA* 'PROFESSOR RONTGEN'
AND *H.* 'CRYSTAL PALACE'

# FLOWERING BULBS
## FOR SUMMER AND FALL

With the arrival of spring, it's time to plant summer- and fall-blooming bulbs that thrive despite the heat and humidity that we can count on in the South.

Before you plant, though, check the specific requirements for individual bulbs. For example, Caladiums need warm soil temperatures to grow (65 to 70 degrees F. at a depth of 6 inches). Once established, Caladium, Elephant Ear, and Canna offer lush and lasting color that carry us through the dog days of summer. Other delights include the sweet scents of Lilies like *Lilium* 'Casa Blanca', and later in August or September, we can look forward to Butterfly Ginger, a real nose pleaser. As late summer rains trigger the magical Rain Lilies, *Zephyranthes candida*, and Surprise Lilies, *Lycoris* species, they remind us that fall is just around the corner.

For the greatest success, plant your bulbs in well-drained soil that has been amended with organic matter. Some bulbs like Elephant Ear require constant moisture to thrive and attain their maximum size.

Whether you incorporate these summer and fall bulbs into the border or grow them in pots, they offer an easy way to add drama and color to your garden with both foliage and flowers.

# UPRIGHT ELEPHANT EAR

*Alocasia* spp. and cultivars

For pizzaz in the summer garden, large Elephant Ear plants are perfect. Tropical favorite 'Frydek' has large, striking, velvety dark green to almost black, downward-facing leaves with bright ivory veins on 1- to 2-foot stems. The leaves become even more dramatic when they are wet. This critter-resistant beauty offers months of color and movement when the leaves sway in the breeze. Grow them in large containers or in the ground as a focal point. Upright Elephant Ear thrives in our summer heat and humidity.

**Size:** The leaves grow up to 18 inches long and 10 inches wide. Mature plants grow 4 to 5 feet tall in no time.

**Conditions:** Shade or part shade is best, but it tolerates sun. Soil that is rich in organic material and will hold moisture is great, but 'Frydek' grows in a wide range of soils including clay or bog. Plant once soil temperatures have warmed.

**Zones:** 8 to 11

**Uses and Companions:** Plant Upright Elephant Ear in large containers alone or with other plants like white Impatiens, red or orange New Guinea Impatiens, or Caladiums. Site it at the edge of a pond or in a bog.

# CALADIUM

*Caladium × hortulanum* cultivars

For color in the shade garden from summer to frost, Caladiums are indispensable. There foliage comes in a wide range of colors and combinations from white and green to deep red and pink. Native to South America, they

thrive in our summer heat and look great both in the garden and in containers. Planted on their own or in combination with other annuals and perennials, they offer months of beauty. There are selections, too, that have been developed to withstand some summer sun.

**Size:** Depending on the selection, some are small and others get quite large. 'Aaron' has medium-sized leaves and grows 18 to 20 inches tall with a 12- to 18-inch spread, and 'Red Flash' grows 15 to 22 inches tall.

**Conditions:** Wait to plant tubers in the ground until soil temperatures are 65 to 70 degrees F. If you plant tubers too early, they may rot. Moist, well-drained soil and part to full shade is ideal.

**Zones:** 9 to 11

**Uses and Companions:** Grow Caladiums on their own or combine them with New Guinea Impatiens, Ferns, and Hostas.

# CANNA

*Canna* spp. and cultivars

Cannas were originally planted for their foliage, not their colorful flowers. Today gardeners don't have to choose; they can have both if they like: large, dramatic leaves and brilliant-colored blooms. When other flowers wilt in the heat of summer, Cannas come to life. Some, like Canna 'Australia', offer dark, almost black leaves and orange-red flowers all summer long. Growing 4 to 5 feet tall, this Canna adds drama to the summer border or next to a pond. It's perfect for a backdrop or in the middle of the garden as a focal point.

**Size:** Cannas grow to 2 to 10 feet tall and 1 to 3 feet wide depending on the selection. Blooms vary from 2 to 5 inches across. The leaves are lance-shaped; some are wider than others.

**Conditions:** Full sun and lots of moisture are ideal, but Cannas survive in drier soils. They can even be grown in pots submerged in water. If leaves become damaged, cut them back to encourage new growth.

**Zones:** 7 to 10

**Uses and Companions:** Site Cannas at the back of the summer border. Combine them with Iris, *Lobelia cardinalis*, Coneflowers, Daylilies, and Elephant Ear for a tropical look.

# ELEPHANT EAR

*Colocasia esculenta* 'Illustris'

If you want to add a tropical flair to your garden, Elephant's Ear 'Illustris', with its heart-shaped black leaves and bright green veins, makes a bold statement. The large leaves dance in the breeze, adding motion and color to the summer garden. The more sun it receives, the deeper purple (almost black with a green background) the leaves turn. However, if you grow 'Illustris' in deep shade, the leaves will turn almost completely green.

**Size:** Elephant's Ear 'Illustris' grows 4 feet tall with leaves that are 2 feet long and 1 foot across.

**Conditions:** Plant Elephants Ear in full sun, part shade, or full shade. Elephants Ear will grow in water but it doesn't like water sitting around its stems, so don't mulch heavily. 'Illustris' is a heavy feeder; use half-strength water-soluble food daily.

**Zones:** 7b to 10

**Uses and Companions:** Site Elephants Ear it at the edge of a pond, in a large pot with other tropicals like purple-leaf Pepper Plants and Coleus, or in the perennial border with Hardy Palms, Cannas, and Bananas.

# CRINUM
*Crinum* cultivars

*Crinum*, an old-fashioned favorite with staying power, offers big and bold straplike leaves and fragrant, lily-shaped flowers from summer through fall. *Crinum* is a prolific bloomer, sending up tall stalks of 4- to 6-inch flowers for weeks. This perennial tuber forms a large mound of foliage and adds color and scent to the summer and fall garden.

**Size:** Crinum Lilies grow to 3 to 5 feet tall by 3 to 5 feet wide.

**Conditions:** This plant prefers full sun or part shade. When you plant this tuber in your garden, select the site carefully, as Crinums can get quite large and don't like to be moved or divided often. Give them a soil that is rich in organic matter.

**Zones:** 7b to 10

**Uses and Companions:** Plant Crinum Lilies near the edge of a pond with Cannas, Elephant's Ear, or on their own. Plant a group as a focal point. Combine it with *Fatsia japonica* for a contrast in foliage.

# BUTTERFLY GINGER
*Hedychium coronarium*

For fragrant flowers in late summer to early fall, Butterfly Ginger is hard to beat. The white flowers look like butterflies or exotic orchids atop tall plants with large

green leaves. Undemanding and easy to grow, this old-fashioned beauty gives a lot and demands little in return. This plant grows from tuberous roots and is easy to divide if clumps get too large.

**Size:** Butterfly Ginger matures at 5 to 6 feet tall and 2 to 3 feet wide.

**Conditions:** Full sun or part shade is best. This Hardy Ginger will tolerate a range of soil types, including those that are compacted and damp.

**Zones:** 7 or 8 to 10

**Uses and Companions:** Plant this at the back of the shrub or perennial border where its foliage will provide a strong vertical accent and its sweet perfume will fill the air. Combine it with other Gingers and late-blooming shrubs and perennials like *Hydrangea paniculata* 'Tardiva' or Butterfly Bush.

# CASA BLANCA ORIENTAL LILY

*Lilium 'Casa Blanca'*

Oriental Lilies add elegance to the summer border, and 'Casa Blanca' is regal with its pure white fragrant flowers in late summer, usually July. The stems grow to a height of 4 feet or taller and the blooms are up to 10 inches across. A showstopper, this sturdy flower also makes a good cut flower and is popular for arrangements. Just a few will add height and drama to your perennial garden.

**Size:** 'Casa Blanca' grows 4 to 5 feet tall and clumps up to 2 feet across.

**Conditions:** Full sun and moist, well-drained soil are ideal for this bulb. Plant 8 inches deep and 6 inches apart in soil that has been amended with organic matter or compost. Fertilize at planting and in early spring. Stake them when they are about half their mature size.

**Zones:** 5 to 9

**Uses and Companions:** Plant Casa Blanca Lilies in the perennial garden with herbaceous plants like *Phlox paniculata* 'David', with white flowers, and *Hydrangea paniculata* 'Tardiva'. Underplant with annuals or Calamint.

Spider Lily is a bulb that shows up in old, established Southern gardens where it has been grown for generations. Clusters (ten to twenty per stem) of bright red, spidery flowers appear in early fall on naked stems when few other bulbs are in bloom. They make cozy companions for Ferns, like Southern Shield Fern, *Thelypteris kunthii,* and other shade-loving perennials such as *Hosta* and Hellebores. Some keep their grassy foliage all winter, and others do not. All are pest resistant and persistent once established.

# SPIDER LILY

*Lycoris radiata*

**Size:** Spider Lily grows 12 to 18 inches tall and 6 to 9 inches wide.

**Conditions:** Part sun, part shade, or half shade and moist, well-drained soil are best. Keep them out of windy areas so that flowers last longer. Plant as soon as you receive bulbs in late May or June. Fertilize with a diluted solution when you plant.

**Zones:** 7 to 10

**Uses and Companions:** If you inherit a bed of English Ivy, plant *Lycoris* with the Ivy for a spectacular display in early autumn. Plant *Lycoris* at the edge of the woodland with Hostas, Hellebores, and Ferns.

# SURPRISE LILY

*Lycoris squamigera*

In late July when the annuals begin to fade, we are surprised when suddenly clumps of rosy pink flowers resembling Amaryllis shoot out of the ground, almost like magic. In spring the wide, gray straplike foliage appears when the Daffodils are in bloom but fades long before the flowers appear.

**Size:** Surprise Lily stalks grow 2½ feet tall with flowers that are 2 to 3 inches across.

**Conditions:** Plant bulbs in late May or June, as soon as you receive them, in moist, well-drained soil. Surprise Lily tolerates a wide range of soil types including sandy and clay. Fertilize in early spring when foliage appears and plants are actively growing.

**Zones:** 5 to 10

**Uses and Companions:** Plant Surprise Lily in the herbaceous border in combination with annuals and perennials. Combine it with the blue flowers of Plumbago or the hardy *Geranium* 'Rozanne'. Plant masses of it against a backdrop of dark green conifers.

# RAIN LILY

*Zephyranthes candida*

Aptly named, Rain Lilies are prolific bloomers. The flowers appear just after a rain, usually in August to September. They look like giant white Crocus and are not fussy about where you plant them. These gems will even grow in pots in the water. The grassy foliage is also attractive. As are other members of the Amaryllis family, they are pest proof, which means critters should leave them alone in your garden.

**Size:** Rain Lily displays 1-inch-tall flowers appearing on 6- to 12-inch-tall stems.

**Conditions:** Full sun and moist, well-drained soil are ideal, but these bulbs will grow in a wide range of soil types.

**Zones:** 7 to 10

**Uses and Companions:** Plant this Rain Lily as an edger for the perennial border or along paths where you can appreciate the flowers. Tuck it into the border between perennials that bloom in the spring so you will be surprised in summer. Pair this bulb with other types of Rain Lilies or with *Lycoris*.

CALLA LILY

# MORE BULBS
## FOR SUMMER AND FALL

*Agapanthus* Headbourne hybrids—*African Lily*

*Alstroemeria psittacina*—*Parrot Lily*

*Crocus speciosus*—*Fall Blooming Crocus*

*Dahlia* spp. and cultivars—*Dahlia*

*Gladiolus callianthus* 'Murielae'—*Abyssinian Gladiolus*

*Lilium formosanum*—*Philippine Lily*

*Hymenocallis coronaria*—*Spider Lily*

*Rhodophiala bifida*—*Oxblood Lily*

*Sprekelia formosissima*—*Aztec Lily*

*Zantedeschia* cultivars—*Calla Lily*

# CREATING DRAMA

There is something about big, bold foliage that appeals to gardeners. Combining tropical looking plants with woody ornamentals can evoke the feeling of an exotic jungle landscape or add a sense of drama to the scene. Plants with oversized leaves include the Umbrella Plant, Banana, Canna, Elephant Ear, Castor Bean, and the Chinese Parasol Tree.

Look to the following plants to add drama to your garden:

BEAR'S BREECH

*Firmiana simplex*, Chinese Parasol Tree—This tree has huge leaves that are 10 to 24 inches across. The bark is green and stands out in the garden. For a contrast, underplant it with Ferns.

*Magnolia macrophylla*, Big Leaf Magnolia and *Acanthus* 'Summer Beauty', Bear's Breech—The dramatic, oversized Magnolia leaves and large, fragrant, white flowers provide the perfect canopy for large clumps of Bear's Breech with its dark, glossy dissected leaves and tall spikes of white flowers marked with purple.

*Musa basjoo*, Hardy Banana, and *Canna indica* 'Red Stripe'—This dramatic combination features a Canna with stalks that are 8 feet tall and purple-and-green leaves that are nearly 2 feet long with a Banana that grows 10 feet tall with leaves that are 6 feet long.

*Ricinus communis*, Castor Bean—This annual grows 6 to 15 feet tall and has leaves that are 1 to 3 feet across. The selection 'Sanguineus' has blood-red foliage. Combine it with colorful annuals.

# DIRECTORY OF SOUTHERN GARDENS

Botanical Gardens and Arboreta (singular is arboretum) grow and cultivate plants, usually for scientific and educational purposes. Most are open to the public for education, research, display, and relaxation. They provide gardeners of all types an opportunity to view plants in a garden setting that they may not otherwise be familiar with. Visit your local public garden and discover the wonders of plants that grow and thrive in your region. For more information go to www.publicgardens .org. This is a partial list of some of the wonderful gardens to consider visiting in the South.

## ALABAMA

Birmingham Botanical Gardens
2612 Lane Park Rd.
Birmingham, AL 35223
(205) 414–3950
www.bbgardens.org

Donald E. Davis Arboretum Biological Sciences
101 Life Sciences Bldg.
Auburn University, AL 36849
(334) 844–5770
www.auburn.edu/arboretum/about-us

Huntsville Botanical Garden
4747 Bob Wallace Ave.
Hunstville, AL 35805
(256) 830–4447
www.hsvbg.org

Mobile Botanical Gardens
P.O. Box 8382
5151 Museum Drive
Mobile, AL 36608
(251) 342–0555
www.mobilebotanicalgardens.org

## ARKANSAS

Botanical Garden of the Ozarks
4703 N. Crossover Road
Fayetteville, AR 72764
(479) 750–2620
www.bgozarks.org

Garvan Woodland Gardens
550 Arkridge Road
P.O. Box 22240
Hot Springs National Park, AR 71913
(800) 366–4664

## GEORGIA

Atlanta Botanical Garden
1345 Piedmont Ave. NE
Atlanta, GA 30309
(404) 876–5859
www.atlantabotanicalgarden.org

Callaway Gardens
P.O. Box 2000
Pine Mountain, GA 31822
(800) 225–5292
www.callawaygardens.org

Georgia Southern Botanical Garden
Georgia Southern University
P.O. Box 8039
Statesboro, GA 30460
(912) 871–1149
welcome.georgiasouthern.edu/garden

Waddell Barnes Botanical Gardens
Macon State College
100 College Station Dr.
Macon, GA 31206
(478) 471–2780
www.maconstate.edu/botanical/
default.aspx

# LOUISIANA
The New Orleans Botanical Garden
City Park
1 Palm Drive
New Orleans, LA 70124
(504) 482-4888
www.neworleanscitypark.com/nobg.
html

Hilltop Arboretum
11855 Highland Road
Baton Rouge, LA 70810
(225) 767–6916
www.lsu.edu/hilltop

# MISSISSIPPI
The Crosby Arboretum
370 Ridge Rd.
Picayune, MS 39644
(601) 799–2311

# NORTH CAROLINA
Cape Fear Botanical Garden
536 N. Easter Blvd.
Fayetteville, NC 28301
(910) 486–0221
info@capefearbg.org

Daniel Stowe Botanical Garden
6500 S. New Hope Rd.
Belmont, NC 28012
(704) 825–4490
www.dsbg.org

JC Raulston Arboretum
NC State University
Dept of Hort. Science
Campus Box 7522
Raliegh, NC 27695-7522
(919) 515–3132
www.ncsu.edu/jcraulstonarboretum/
index.php

North Carolina Botanical Garden
University of North Carolina at
Chapel Hill
CB 3375
Totten Center
Chapel Hill, NC 27599
(919) 962–0522
ncbg@unc.edu

The North Carolina Arboretum
100 Frederick Law Olmsted Way
Asheville, NC 28806
(828) 665–2492
www.ncarboretum.org

# SOUTH CAROLINA
Brookgreen Gardens
1931 Brookgreen Dr.
Murrells Inlet, SC 29576 (Pawleys
Island, SC )
(843) 235–6000
www.brookgreengarden.org

South Carolina Botanical Garden
Clemson University
Public Service & Agriculture
130 Lehotsky Hall
Clemson, SC 29634
(864) 656–3015
www.virtual.clemson.edu

# TENNESSEE

Cheekwood Botanical Garden
and Museum of Art
1200 Forrest Park Drive
Nashville, TN 37205
(615) 356–8000
www.cheekwood.org/gardens

University of Tennessee
Arboretum—Oak Ridge
901 South Illinois Ave.
Oak Ridge, TN 37830
(865) 483–3571
www.discover.org/utarboretum

# TEXAS

Clark Gardens Botanical Park
567 Maddux Rd.
Weatherford, TX 76068
(940) 682–4856
www.clarkgardens.com

Dallas Arboretum &
Botanical Society, Inc.
8525 Garland Rd.
Dallas, TX 75218
(214) 515–6500
www.dallasarboretum.org

Fort Worth Botanic Garden
3220 Botanic Garden Blvd.
Fort Worth, TX 76107
(817) 871–7686
www.fwbg.org

Houston Arboretum & Nature Center
4501 Woodway Drive
Houston, TX 77024
(713) 681–8433
www.houstonarboretum.org

Mercer Arboretum & Botanic Gardens
22306 Aldine Westfield Rd.
Humble, TX 77338
(281) 443–8731
www.hcp4.net/mercer

San Antonio Botanical Garden
555 Funston Place
(North New Braunfels Ave.)
San Antonio, TX 78209
(210) 207–3250
www.sabot.org

# BIBLIOGRAPHY

No good gardening book is complete without references. The following books have been invaluable to me, and we invite you to discover the tremendous amount of horticultural information contained in them.

Armitage, Allan. *Herbaceous Perennial Plants: A Treatise on Their Identification, Culture, and Garden Attributes*. Champaign, IL: Stipes Publishing, 1997.

Armitage, Allan. *Armitage's Garden Perennials: A Color Encyclopedia*. Portland, OR: Timber Press, 2000.

Armitage, Allan. *Armitage's Manual of Annuals, Biennials, and Half-Hardy Perennials*. Champaign, IL: Stipes Publishing, 2001.

Armitage, Allan. *Herbaceous Perennial Plants: A Treatise on Their Identification, Culture, and Garden Attributes, Third Edition*. Champaign, IL: Stipes Publishing, 2008.

Bender, Steve, and Felder Rushing. *Passalong Plants*. Chapel Hill, NC: The University of North Carolina Press, 1993.

Brooklyn Botanic Garden. Plants and Gardens Handbooks. Many different subjects are available; for a list, contact the Brooklyn Botanic Garden, 1000 Washington Ave., Brooklyn, NY.

Burke, Ken (ed.). *Shrubs and Hedges*. Franklin Center, PA: The American Horticultural Society, 1980.

Burke, Ken (ed.). *Gardening in the Shade*. Franklin Center, PA: The American Horticultural Society, 1982.

Copeland, Linda, and Allan Armitage. *Legends in the Garden: Who in the World Is Nellie Stevens?* Atlanta, GA: Wings Publishers, 2001.

Dirr, Michael. *Manual of Woody Landscape Plants*. Champaign, IL: Stipes Publishing, 1998.

Gardiner, J.M. *Magnolias*. Chester, PA: Globe Pequot Press, 1989.

Gates, Galen, et al. *Shrubs and Vines*. New York, NY: Pantheon Books, 1994.

Glasener, Erica, and Walter Reeves. *Georgia Gardener's Guide, Revised Edition*. Franklin, TN: Cool Springs Press, 2004.

Glasener, Erica, and Walter Reeves. *Month-By-Month™ Gardening in Georgia, Revised Edition*. Franklin, TN: Cool Springs Press, 2006.

Greenlee, John. *The Encyclopedia of Ornamental Grasses*. Emmaus, PA: Rodale Press, 1992.

Halfacre, R. Gordon, and Anne R. Shawcroft. *Landscape Plants of the Southeast*. Raleigh, NC: Sparks Press, 1979.

Harper, Pamela. *Time-Tested Plants: Thirty Years in a Four-Season Garden*. Portland, OR: Timber Press, 2000.

Harper, Pamela, and Frederick McGourty. *Perennials: How to Select, Grow and Enjoy*. Tucson, AZ: HP Books, 1985.

Heath, Brent, and Becky. *Daffodils for American Gardens*. Washington, D.C.: Elliott & Clark Publishing, 1995.

Hipps, Carol Bishop. *In a Southern Garden*. New York, NY: Macmillan Publishing, 1994.

Lawrence, Elizabeth. *A Southern Garden*. Chapel Hill, NC: The University of North Carolina Press, 1991.

Lawson-Hall, Toni, and Brian Rothera. *Hydrangeas*. Portland, OR: Timber Press, 1996.

Loewer, Peter. *Tough Plants for Tough Places*. Emmaus, PA: Rodale Press, 1992.

MacKenzie, David. *Perennial Groundcovers*. Portland, OR: Timber Press, 1997.

Mikel, John. *Ferns for American Gardens*. New York, NY: Macmillan Publishing, 1994.

Ogden, Scott. *Garden Bulbs for the South*. Dallas, TX: Taylor Publishing, 1994.

Overy, Angela. *Sex in Your Garden*. Golden, CO: Fulcrum Publishing, 1997.

Still, Steven. *Manual of Herbaceous Ornamental Plants, 4th edition*. Champaign, IL: Stipes Publishing, 1994.

Vengris, Jonas, and William A. Torello. *Lawns*. Fresno, CA: Thomson Publications, 1982.

Wilson, Jim. *Bullet-Proof Flowers for the South*. Dallas, TX: Taylor Publishing, 1999.

Winterrowd, Wayne. *Annuals for Connoisseurs*. New York, NY: Prentice Hall, 1992.

**Alkaline soil**: soil with a pH greater than 7.0. It lacks acidity, often because it has limestone in it.

**All-purpose fertilizer**: powdered, liquid, or granular fertilizer with a balanced proportion of the three key nutrients—nitrogen (N), phosphorus (P), and potassium (K). It is suitable for maintenance nutrition for most plants.

**Annual**: a plant that lives its entire life in one season. It is genetically determined to germinate, grow, flower, set seed, and die the same year.

**Balled and burlapped**: describes a tree or shrub grown in the field whose rootball was wrapped with protective burlap and twine when the plant was dug up to be sold or transplanted.

**Bare root**: describes plants that have been packaged without any soil around their roots. (Often young shrubs and trees purchased through the mail arrive with their exposed roots covered with moist peat or sphagnum moss, sawdust, or similar material, and wrapped in plastic.)

**Barrier plant**: a plant that has intimidating thorns or spines and is sited purposely to block foot traffic or other access to the home or yard.

**Beneficial insects**: insects or their larvae that prey on pest organisms and their eggs. They may be flying insects, such as ladybugs, parasitic wasps, praying mantids, and soldier bugs, or soil dwellers such as predatory nematodes, spiders, and ants.

**Berm**: a narrow, raised ring of soil around a tree, used to hold water so it will be directed to the root zone.

**Bract**: a modified leaf structure on a plant stem near its flower that resembles a petal. Often it is more colorful and visible than the actual flower, as on a dogwood.

**Bud union**: the place where the top of a plant was grafted to the rootstock; usually refers to roses.

**Canopy**: the overhead branching area of a tree, usually referring to its extent including foliage.

**Cold hardiness**: the ability of a perennial plant to survive the winter cold in a particular area.

**Composite**: a flower that is actually composed of many tiny flowers. Typically, they are flat clusters of disklike flowers, sometimes surrounded by ray-petaled flowers. Composite flowers are highly attractive to bees and beneficial insects.

**Compost**: organic matter that has undergone progressive decomposition until it is reduced to a spongy, fluffy texture. Added to soil of any type, it improves the soil's ability to hold air and water and to drain well.

**Corm**: the swollen energy-storing structure, analogous to a bulb, under the soil at the base of the stem of plants such as crocus and gladiolus.

**Crown**: the base of a plant at, or just beneath, the surface of the soil where the roots meet the stems.

**Cultivar**: a **CULTI**vated **VAR**iety. It is a naturally occurring form of a plant that has been identified as special or superior and is purposely selected for propagation and production.

**Deadhead**: a pruning technique that removes faded flower heads from plants to improve their appearance, abort seed production, and stimulate further flowering.

**Deciduous plants**: unlike evergreens, these trees and shrubs lose their leaves in the fall.

**Desiccation**: drying out of foliage tissues, usually due to drought or wind.

**Division**: the practice of splitting apart perennial plants to create several smaller-rooted segments. The practice is useful for controlling the plant's size and for acquiring more plants; it is also essential to the health and continued flowering of certain ones.

**Dormancy**: the period, usually the winter, when perennial plants temporarily cease active growth and rest. Dormant is the verb form, as used in this sentence: Some plants, like spring-blooming bulbs, go dormant in the summer.

**Established**: the point at which a newly planted tree, shrub, or flower begins to produce new growth, either foliage or stems. This is an indication that the roots have recovered from transplant shock and have begun to grow and spread.

**Evergreen**: perennial plants that do not lose their foliage annually with the onset of winter. Needled or broadleaf foliage persists and continues to function on a plant through one or more winters, aging and dropping unobtrusively in cycles of three or four years or more.

**Foliar**: of or about foliage; usually refers to the practice of spraying foliage, as in fertilizing or treating with insecticide; leaf tissues absorb liquid directly for fast results.

**Floret**: a tiny flower, usually one of many forming a cluster, that comprises a single blossom.

**Germinate**: to sprout. Germination is a fertile seed's first stage of development.

**Graft (union)**: the point on the stem of a woody plant with vigorous roots where a stem from a plant is inserted so that it will join with it. Roses are commonly grafted.

**Hardscape**: the permanent, structural, nonplant part of a landscape, such as walls, sheds, pools, patios, arbors, and walkways.

**Herbaceous**: plants having fleshy or soft stems that die back with frost; the opposite of woody.

**Hybrid**: a plant that is the result of intentional or natural cross-pollination between two or more plants of the same species or genus.

**Low water demand**: describes plants that tolerate dry soil for varying periods of time. Typically, they have succulent, hairy, or silvery-gray foliage and tuberous roots or taproots.

**Mulch**: a layer of material over bare soil to protect it from erosion and compaction by rain, and to inhibit weeds. It may be inorganic (gravel, fabric) or organic (wood chips, bark, pine needles, chopped leaves).

**Naturalize**: (a) to plant seeds, bulbs, or plants in a random, informal pattern as they would appear in their natural habitat; (b) to adapt to and spread throughout adopted habitats (a tendency of some nonnative plants).

**Nectar**: the sweet fluid produced by glands on flowers that attract pollinators such as hummingbirds and honeybees for whom it is a source of energy.

**Nonselective weedkiller**: a herbicide that kills every green plant it touches. Examples: glyphosate (Roundup®), glufosinate (Finale®), diquat, and vinegar.

**Organic matter**: any material or debris that is derived from plants. It is carbon-based material capable of undergoing decomposition and decay.

**Peat moss**: organic matter from peat sedges (United States) or sphagnum mosses (Canada), often used to improve soil texture. The acidity of sphagnum peat moss makes it ideal for boosting or maintaining soil acidity while also improving its drainage.

**Perennial**: a flowering plant that lives over two or more seasons. Many die back with frost, but their roots survive the winter and generate new shoots in the spring.

**pH**: a measurement of the relative acidity (low pH) or alkalinity (high pH) of soil (or water) based on a scale of 1 to 14, 7 being neutral. Individual plants have a pH range in which they grow best.

**Pinch**: to remove tender stems and/or leaves by pressing them between thumb and forefinger. This pruning technique encourages branching, compactness, and flowering in plants, or it removes aphids clustered at growing tips.

**Pollen**: the yellow, powdery grains in the center of a flower. A plant's male sex cells are transferred to the female plant parts by means of wind, insect, or animal pollinators to fertilize them and create seeds.

**Raceme**: an arrangement of single-stalked flowers along an elongated, unbranched axis.

**Rhizome**: a swollen energy-storing stem structure, similar to a bulb, that lies horizontally in the soil, with roots emerging from its lower surface and growth shoots from a growing point at or near its tip, as in bearded iris.

**Rootbound (or potbound)**: the condition of a plant that has been confined in a container too long, its roots having been forced to wrap around themselves and even swell out of the container. Successful transplanting or repotting requires untangling and trimming away of some of the matted roots.

**Root flare**: the transition at the base of a tree trunk where the bark tissue begins to differentiate and roots begin to form just before entering the soil. This area should not be covered with soil when planting a tree.

**Self-seeding**: the tendency of some plants to sow their seeds freely around the yard. It creates many seedlings the following season that may or may not be welcome.

**Semievergreen**: tending to be evergreen in a mild climate but deciduous in a rigorous one.

**Shearing**: the pruning technique whereby plant stems and branches are cut uniformly with long-bladed pruning shears (hedge shears) or powered hedge trimmers. It is used when creating and maintaining hedges and topiary.

**Slow-release fertilizer**: fertilizer that is water insoluble and therefore releases its nutrients gradually as a function of soil temperature, moisture, and related microbial activity. Typically granular, it may be organic or synthetic.

**spp.**: a commonly used abbreviation for the word species.

**Succulent growth**: the sometimes undesirable production of fleshy, water-storing leaves or stems that results from overfertilization.

**Sucker**: a new growing shoot. Underground plant roots produce suckers to form new stems and spread by means of these suckering roots to form large plantings, or colonies. Some plants produce root suckers or branch suckers as a result of pruning or wounding.

**Tuber**: a type of underground storage structure in a plant stem, analogous to a bulb. It generates roots below and stems above ground (example: dahlia).

**Variegated**: having various colors or color patterns. The term usually refers to plant foliage that is streaked, edged, blotched, or mottled with a contrasting color, often green with yellow, cream, or white.

**White grubs**: fat, off-white, wormlike larvae of summer beetles. They reside in the soil and feed on plant (especially grass) roots until summer when they emerge as beetles to feed on plant foliage.

**Wings**: (a) the corky tissue that forms edges along the twigs of some woody plants such as winged euonymus; (b) the flat, dried extension of tissue on some seeds, such as maple, that catch the wind and help them disseminate.

# PHOTOGRAPHY CREDITS

The photographs in this book are here to inform and inspire you.

**Erica Glasener:** 5, 20a, 21, 22, 23, 24, 27, 30b, 32, 33a, 40, 41a, 42, 43, 44a, 50, 51b, 52, 54a, 62b, 63, 70a, 71b, 73, 74b, 80, 82b, 83b, 87, 90b, 92, 94, 96, 100b, 101a, 102b, 103a, 104, 105, 106, 110a, 111a, 112a, 113, 120b, 121b, 122, 123a, 124b, 125, 126, 127, 131b, 132a, 134a, 137, 140a, 141b, 143a, 144a, 145, 150a, 151b, 154, 171b, 174a, 180a, 181a, 183, 184, 190, 191, 192a, 194, 201, 202b, 203, 204, 207, 210, 211, 212, 213a, 214b

---

**William Adams:** 20b, 153a, 167
**Jennifer Anderson:** 86
**Liz Ball:** 12, 53b, 55, 82a, 83a, 162b, 164b, 173a, 187, 200b
**Liz Ball and Rick Ray:** 62a , 102a, 142a, 150b, 171a, 173b, 202a
**K. Bussolini:** 15
**Lana Coit:** 155, 172
**Mike Dirr:** 161a, 172a, 192b, 213b
**Thomas Eltzroth:** 24, 33b, 34b, 44b, 46, 47, 51a, 61a, 65, 70b, 71a, 75, 76, 77, 81b, 90a, 91a, 93a, 95, 97, 101b, 103b, 111b, 117, 123b, 124a, 131a, 132b, 133a, 134b, 135a, 143b, 144b, 147, 151a, 161b, 162a, 163, 174b, 176, 182a, 193a, 195, 196, 214a, 215
**Derek Fell:** 182b
**Lou Freeman:** 238
**Pam Harper:** 64b, 72a, 81a, 91b, 100a, 115, 121a, 140b, 164a, 165
**Jupiter Images:** 30, 136
**Dency Kane:** 142b, 152b
**Bill Kersey Graphics:** 14
**Johansen Krause:** 133b, 193b
**Peter Loewer:** 85
**Robert Lyons:** 170b
**Charles Mann:** 45, 216
**J. Mielke:** 197
**Tom McKay:** 53a
**Jerry Pavia:** 25, 34a, 37, 41b, 57, 60, 61b, 64a, 67, 72b, 93b, 107, 112b, 114, 130b, 141a, 152a, 160b, 170a, 175, 180b, 181b, 185, 205
**Marc Pewitt:** 31a
**Carol Reese:** 36
**Mary Robson:** 146
**Felder Rushing:** 26, 35, 66, 74a, 116, 120a, 157, 217
**Neil Soderstrom:** 11, 13, 110b
**Mark Turner:** 130a, 160a, 200a
**André Viette:** 54b, 56
**David Winger:** 10

# ABOUT ERICA

Erica Glasener is nationally known as the host of HGTV's *A Gardener's Diary*. Since 1995 she has introduced her audience to horticulture professionals, specialty plant growers, hobby gardeners, landscape architects, artists and many more in the gardening world through her popular TV show. Erica has traveled across the country to find the most interesting gardeners, to help them tell their stories and showcase their work.

An experienced writer, Erica is the co-author of *Month-By-Month™ Gardening in Georgia* (new edition 2007) and the *Georgia Gardener's Guide* (co-authored with Walter Reeves, revised edition published 2004). She writes an online garden column for *The Atlanta Journal-Constitution* and a bimonthly garden column for *Southern Lady* magazine. As a frequent guest on local, regional and national lifestyle radio programs, she likes to help people solve garden problems.

Erica's horticulture knowledge, coupled with her passion for all things green, has made her a sought-after speaker nationwide at garden and flower shows, garden clubs, home shows, civic organizations and special events. Her ease of connecting with people insures she can educate seasoned gardeners while inspiring those who are just beginning to dig in the dirt.

Putting theory into practice, Erica has designed and now maintains a beautiful garden at her home. Here she grows a wide variety of bulbs, herbs, vegetables, perennials, shrubs and trees. She experiments with new plants and innovative plant groupings, always looking for exciting plant introductions and beautiful color combinations.

Erica earned a degree in Ornamental Horticulture from the University of Maryland. She has served as a contributing editor for *Fine Gardening* and her articles have appeared in *The New York Times*, *The Old Farmer's Almanac*, *Atlanta Magazine* and *Green Guide*.

She lives and gardens with her husband and daughter in Atlanta, Georgia, where she is involved in a number of non-profit organizations, offering hands-on expertise and insight into a variety of horticulture projects and topics.

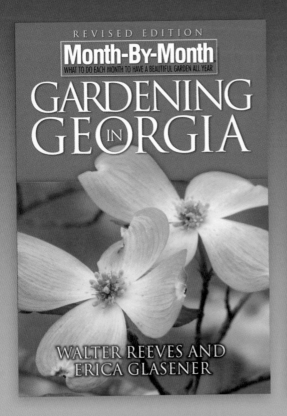
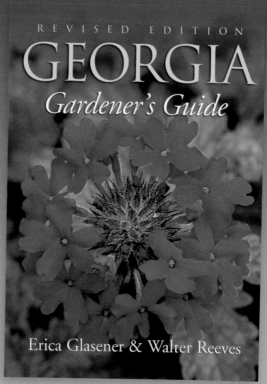